MEDIATED MEMORIES IN THE DIGITAL AGE

Cultural Memory

in

the

Present

Mieke Bal and Hent de Vries, Editors

MEDIATED MEMORIES IN THE DIGITAL AGE

José van Dijck

STANFORD UNIVERSITY PRESS

STANFORD, CALIFORNIA

2007

Stanford University Press
Stanford, California

Printed in the United States of America on acid-free, archival-quality paper

Library of Congress Cataloging-in-Publication Data

Dijck, José van.
   Mediated memories in the digital age / José van Dijck.
      p. cm. — (Cultural memory in the present)
   Includes bibliographical references and index.
   ISBN 978-0-8047-5623-5 (cloth) — ISBN 978-0-8047-5624-2 (paper)
      1. Image processing—Digital techniques   2. Multimedia systems.
3. Memory. I. Title.

TA1637.D57 2007
153.1'3—dc22

                                                                    2007008641

   Typeset by Westchester Book Group in 11/13.5 Garamond

They came to know the incorrigible sorrow of all prisoners and ex-iles, which is to live in company with a memory that serves no pur-pose. Even the past, of which they thought incessantly, had a savor only of regret. . . . Hostile to the past, impatient of the present, and cheated of the future . . . the only way of escaping from that intolerable leisure was to set the trains running again in one's imagination.

—ALBERT CAMUS, *THE PLAGUE,* 1946

Memory must be set to some purpose

Contents

Preface   xi

Acknowledgments   xvii

1. Mediated Memories as a Conceptual Tool   1
2. Memory Matters in the Digital Age   27
3. Writing the Self   53
4. Record and Hold   77
5. Pictures of Life, Living Pictures   98
6. Projecting the Family's Future Past   122
7. From Shoebox to Digital Memory Machine   148
8. Epilogue   170

Notes   183

Bibliography   217

Index   229

Delany - semiotics - representation/translation
of slavery

WPA narratives

memory

Preface

[handwritten annotations: is this a different idea? memory—digital—nostalgia —translating memory into digital— same idea—what gets lost—changed in translation?]

In the summer of 2003, I bought myself a digital camera; not exactly an early adaptor, I had followed the advice of relatives and friends who lauded its creative potential, assuring it would give a new impulse to my sluggish career as a snapshot photographer. Within less than a month, I had taken a couple of hundred pictures, not counting the images I had already removed from the camera or those I had deleted from my hard disk. My acquisition sparked new enthusiasm, particularly when I started to sort out and scan some old laminated pictures that had been tucked away in unsorted batches in several shoeboxes. Photoshop software enabled me to doctor, scan, and recombine old pictures into surprising contexts and opened up new vistas for future camera use. In addition to photographs, hundreds of music files on my hard disk were testimony to a previous digital discovery—the file-sharing system Kazaa—that had unlaced a peculiar craving for old songs, which I then transferred to my MP3 player. These songs, in turn, led me to dig up my almost-forgotten boxes containing handwritten letters, diaries, and notes—a discovery that made me wonder whether I should scan these items into my computer before they were completely unreadable as a result of fading ink. And finally, I had to face the hundreds of old videotapes sitting on the shelves of a large closet, gathering dust after my VCR had broken down, which prompted me to switch to a DVD recorder. The prospect of having to select and transfer hundreds of old music albums, photos, (home) videos, and letters caused a mixture of excitement and weariness. Excitement stirred because of the potential for uninhibited, nostalgic yearning while sorting through these analog items and perhaps the chance to recycle some in the creation of new memorable insights and objects. Weariness set in because I realized that such a rescue-my-past-operation would not only be time consuming, but it also would involve agonizing decisions about what items to store and what to throw away after digitization.

At first, I tackled the problem as a practical one: a mess of objects in need of systematic sorting and filing. When I sat down to think about it, the problem was not the mess in my shoebox and by extension on my desktop; these objects and technologies generated questions concerning memory and media that were far more intricate than I had initially thought. The contents of my shoebox indeed posed the challenge of fitting digital memories into analog frames for storage and retrieval, but beyond my idiosyncratic dilemmas, it raised poignant concerns about the relation between material objects and autobiographical memory, between media technologies and our habits and rituals of remembrance. My collection of personal items made me reflect on how we present and preserve images of ourselves to others; it caused me to speculate on how private collections tap into the much larger phenomenon of communal rites of storing and re-trieving. I wondered whether my switch from analog to digital memory objects was the result of a general technological-commercial push or the desire to be more creatively engaged with my treasured artifacts. The microuniverse of my shoebox opened up a Pandora's box of unexpected philosophical questions pertaining to the nature, culture, and politics of what I dubbed "mediated memories." Why and how do we create and save mediated items for later reminiscence? What is the function of mediated memories in our personal lives? What is the role of media technologies and material objects in capturing both individual and collective memory? Are analog and digital objects interchangeable in the making, storing, and recalling of memories? Do digital objects change our inscription and re-membrance of lived experience, and do they affect the memory process in our brains?

Articulating these questions was a first step toward acknowledging the complexity of the term I had casually coined. "Mediated memories" refers to both to the concrete objects in my shoebox and a mental concept— a concept that encompasses aspects of mind and body as well as of tech-nology and culture. I also realized that memory and media both comprise vast and exhaustively mined research subjects, to the extent that the terms themselves are at risk of becoming empty signifiers. To say that human memory is a complex problem is an understatement; it is such a daunting, intricate object of research that generations of scholars can hardly be expected to map its mechanisms. Academics from a variety of disciplinary backgrounds, from the biomedical sciences to cultural theory, examine how and why we remember and which basic mechanisms scaffold our processes of recollection. Neurobiologists delve into the operational physiology of memory functions as they relate to *human nature.* Basic research into genetic, neurological, and cognitive aspects of the brain helps explain the effects of injuries and disease on different types of memory. A small portion of this research targets autobiographical memory and the role of emotion in personal reminiscing. Autobiographical memory historically has been the province of psychologists, who examine its working as an aspect of *human behavior.* Cognitive philosophers and social constructivists point at the importance of *materiality and technology*—particularly media technologies and objects—when addressing the issue of memory. Perspectives magnifying memory as a feature of *human culture* prevail in the humanities, most notably in history and cultural studies. In recent years, historians have frequently commented upon the role of media in interlinking our past and present, whereas cultural theorists have engaged with questions of identity and collective memory in the wake of major historical changes, such as exile, war, or diaspora.

In facing such a complex subject as human memory, it is helpful to break apart the components of nature, behavior, materiality, and culture and to scrutinize them in isolation. But it is equally useful, at some point, to put them back together again and acknowledge their conjunction. The question of memory ties together the intricacies of the brain with the dynamics of social behavior and the multilayered density of material and social culture. The concrete contents of my shoebox, caught in a limbo between analog and digital materialization, provides a window onto contemporary debates addressing the relation between self and others, material and virtual, private and public, individual and collective. *Mediated Memories* aims to theorize our personal shoeboxes by turning them into a prism— a conceptual tool through which we can understand larger transformations currently at work in our culture. These larger transformations include but are not restricted to the issue of digitization; digitization only partly reveals the complex interconnections between mind, technology, and culture.

Before taking on specific, concrete examples of shoebox contents, I will first sketch the theoretical and paradigmatic scope of this book. The first chapter interrogates the notion of cultural memory and the role of media in its formation. Media are pivotal to the construction of individual and collective identity—creative acts and products through which people make sense of their lives and the lives of others and connect past to future. The integration of media in the construction of memory has urged social scientists and cultural theorists to define the "mediation of memory," a concept that is useful but often proves to be inconsistent when deployed to explain the mutual shaping of individual and collective memory. "Mediated memories," a modified version of the mediation concept, aims to mend its conceptual flaws.

Chapter 2 addresses the "matter" memory is made of. The genetic, cellular, and cognitive dimensions of human memory are favored subjects for neurobiologists and cognitive scientists; recent research has yielded groundbreaking insights in the mutability of autobiographical memory. Philosophers of mind and social constructivists emphasize the importance of memory's material and technological dimensions in addition to its physiological strata. And cultural theorists concentrate on the significance of sociocultural practices in the manifestation of remembrance. In other words, mediated memories are concurrently embodied in the human brain or mind, enabled by technologies and objects, and embedded in social and cultural contexts of their use. Moving from a culture in which analog memory objects (photographs, diaries, home videos, etc.) prevail to a situation where digital objects become the norm, the questions of why and how memory matters become even more poignant. Dimensions of multidisciplinarity and multimediality are etched into the model of mediated memories introduced in Chapter 1 and expanded in Chapter 2; this model enables us to analyze cultural memory in transition.

The next four chapters take on specific types of contents filling our contemporary shoeboxes, each concentrating on a particular sensory mode privileged by succinct analog media. Words, sounds, still images, and moving images constitute the dominant ingredients, respectively, of diaries, audiotapes, photographs, and home movies or videos—analog items that are presently complemented by digital multimedia forms. Chapter 3 focuses on diaries and lifelogs—a type of weblog considered the digital variant of paper diaries. Long regarded as a reflective and strictly private genre, paper diaries always also had a communicative and public function. These days, the vastly popular cultural practice of blogging and the numerous possibilities for publishing lifelogs online bolster the diary's previously understated communicative and public functions. The Internet, with its intrinsic propensity toward sharing and instant communication, seems to undercut and yet enhance traditional diary features. Seen in this light, we can trace how new digital technologies are actually transforming our notions of privacy and openness, but they also cast a different light on the relation between personal memory and lived experience.

The next chapter, "Record and Hold," deals with the audio component of mediated memories. Recorded popular music is vital to the construction of personal memory and individual identity as well as to the formation of collective memory and musical heritage, a heritage that has grown over time and continues to evolve. Chapter 4 examines the role of popular music in the formation of individual remembering and collective heritage. Remembering through music is more than just an individual act: we need public spaces to share narratives and build a creative commons.

Chapters 5 and 6 attend to visual modes of perception, as embodied by instruments for capturing individual lives in still and moving images. Photography has probably been the favorite medium for arresting life in its formative moments—a vain attempt to freeze time as the future unfolds. Besides its traditional function to confine the past in pictures, digital cameras are now also deployed to communicate in the language of photography and transform the act of memory into an act of experience. The malleability of memory also forms the focus of Chapter 6, but whereas the previous chapter focuses on still images of individuals, this one highlights how family life is captured in moving images. Generations of instruments (8 mm movies, video, digital video, and webcams) have delivered moving pictures of changing mores in generations of families. While digital equipment allows the skewering of diverse historical home modes, it also divulges its inherent technological ability to shape and manipulate memory.

Finally, after reviewing four concrete instances of transforming mediated memories, I return to the problem of storing and retrieving items from the shoebox and desktop. The search for an all-encompassing, universal memory machine is not exactly new; Chapter 7 recounts how memory machines have been the focus of historical and contemporary endeavors to tackle the problem of information overload in the context of personal lives. Computer scientists and engineers have proposed a range of technovisionary solutions and also embroider on historical utopias by designing new fantastic projects that supposedly meet our desire for a comprehensive memory apparatus. I conclude by sketching how digitization, multimedia-tization, and "googlization" may redefine memory, the performance of which was once consigned to the brain or, in contrast, boarded out to the machine.

In these seven chapters, I hope to provide a better understanding of cultural memory, how and why it matters, and what it means in an era of technological, social, and cultural transformation. What started out as a practical problem—the reorganization of my private shoebox filled with memory objects—gradually turned into an academic investigation, spark-ing profound epistemological, ontological, and pragmatic questions and resulting in the design of a theoretical model that helps analyze concrete instances of cultural memory. I have no illusion that this book provides ul-timate answers to all these questions, but it is a modest proposal to rear티c-ulate the changing meaning of cultural memory at a time of transition and bring together a number of diverging disciplinary perspectives to open up new outlooks on this fascinating subject.

Acknowledgments

By the time one writes the acknowledgments, a book has almost become a material artifact, and the writing process is about to become a memory. But the ultimate memory product is full of traces of collaboration, inspiration, and affection. I would like to extend my gratefulness to some of the people who have contributed to this book.

Students and colleagues at the Department of Media Studies at the University of Amsterdam have offered invaluable support and stimuli. My graduate students always shared my enthusiasm for this topic, and if not, they never showed it. Thanks to my dear colleagues Thomas Elsaesser, Frank van Vree, and Eric Ketelaar for co-teaching a seminar on media and memory. Patricia Pisters subtly mended the Deleuze-gaps in my education and has proven a wonderful ally. Without the department's support staff, life would be half as memorable: Jobien Kuiper; Piet van Wijk, Henny Bouwmeester, who also provided valuable administrative support; and Joost Bolten, who is a Word whiz. Being the chair of such a great team of faculty and staff has been an honor and a total delight; the department's growth and flourishing in the past five years was the accomplishment of a superb group.

Some chapters in this book have been the outcome of or input for various projects. I thank Karin Bijsterveld for our collaboration in Sound Technologies and Cultural Practice, a project funded by the Netherlands Organization for Scientific Research (NWO). Sonja Neef's organizing qualities in our handwriting project cannot be praised enough. American colleagues, most notably Richard Grusin, Hugh Crawford, Jonathan Sterne, Phillip Thurtle, Robert Mitchell, Herman Gray, George Lipsitz, and Lisa Cartwright offered valuable criticism to early drafts or lectures on this topic.

Some chapters in this book have roots in journal articles or collections of articles I have written. I have used parts of the following previously published materials: "Mediated Memories: Personal Cultural Memory as an Object of Cultural Analysis." *Continuum: Journal of Media and Cultural Studies* 18 (2004): 261–277; "Memory Matters in the Digital Age." *Configurations* 12 (2006): 349–373; "Composing the Self: Of Diaries and Lifelogs." *Fibreculture* 3 (2004); "Record and Hold: Popular Music between Personal and Collective Memory." *Critical Studies in Media Communication* 25, no 5 (2006): 357–375; "Digitized Memories: The Computer as Personal Memory Machine." *New Media and Society* 7, no. 2 (2005): 291–312; "Future Memories: The Construction of Cinematic Hindsight." In *Theory, Culture and Society* (forthcoming). I thank all anonymous referees for their constructive comments.

Reviewers and editors at Stanford University Press showed confidence in this project from the beginning; thanks to Norris Pope for his encouragement and Deborah Masi for her editorial support. Mieke Bal has been more than a wonderful series editor; she has also been a longtime role model and intellectual source of inspiration.

California and Santa Cruz provided the geographical subtext to this book, the larger part of which was written in Craig Reinarman's lovely house, providing a second home; the idyllic surroundings of the University of California–Santa Cruz campus and the hospitality of the Sociology Department's faculty in the spring of 2005 substantially contributed to my happiness.

I have long depleted my vocabulary to express gratitude toward Ton, so I will keep it simple this time: I owe you. This book is dedicated to my sister Ria, who witnessed its progress in California but did not live to see it in print. She will be etched forever in my memory, with immense fondness and deep respect.

AMSTERDAM, JANUARY 2, 2007

1

Mediated Memories as
a Conceptual Tool

Many people nurture a shoebox in which they store a variety of items signaling their pasts: photos, albums, letters, diaries, clippings, notes, and so forth. Add audio and video tape recordings to this collection as well as all digital counterparts of these cherished items, and you have what I call "mediated memories." These items mediate not only remembrances of things past; they also mediate relationships between individuals and groups of any kind (such as a family, school classes, and scouting clubs), and they are made by media technologies (everything from pencils and cassette recorders to computers and digital cameras). We commonly cherish our mediated memories as a formative part of our autobiographical and cultural identities; the accumulated items typically reflect the shaping of an individual in a historical time frame. But besides their personal value, collections of mediated memories raise interesting questions about a person's identity in a specific culture at a certain moment in time.

Putting these "shoebox" collections at the center of a theoretical and analytical inquiry, this chapter investigates two questions and one concept. First, what is *personal* cultural memory and how does it relate to collective identity and memory? We can distinguish—though not separate—the construction of autobiographical memory as it is grounded in individual psyches from the social structures and cultural conventions that inform it. Personal (re)collections are often subsumed as building blocks of collective history rather than considered in their own right. Personal *cultural* memory emphasizes the value of items as "mediators" between individuals and collectivity, while concurrently signifying tensions between private and public. The growing importance of media technologies to the construction of personal remembrance gives rise to a second pertinent question: what exactly is the nature of memory's mediation? Media technologies and objects, far from being external instruments for "holding" versions of the past, help constitute a sense of past—both in terms of our private lives and of history at large. Memory and media have both been referred to metaphorically as reservoirs, holding our past experiences and knowledge for future use. But neither memories nor media are passive go-betweens: their mediation intrinsically shapes the way we build up and retain a sense of individuality and community, of identity and history.

Therefore, I introduce the concept of mediated memories not only to account for the intricate connection between personal collections and collectivity but also to help theorize the *mutual shaping* of memory and media. By defining and refining this concept into an analytical tool, I hope to turn the items in our private shoeboxes into valuable objects for cultural analysis. As private collections, mediated memories form sites where the personal and the collective meet, interact, and clash; from these encounters we may derive important cultural knowledge about the construction of historical and contemporaneous selves in the course of time: How do our media tools mold our process of remembering and vice versa? How does remembrance affect the way we deploy media devices?

Personal Cultural Memory

The study of what constitutes personal memory has traditionally been the domain of neuroscientists, psychologists, and cognitive theorists. We commonly think of memory as something we have or lack; studies of memory are concerned with our ability to remember or our proclivity to forget things. The majority of studies on memory in the area of psychology deal with our cognitive abilities for recall, and out of those studies, a fair number concentrate on autobiographical or personal memory.[1] The interconnection of memory and self, psychologists state, is crucial to any human being's development. Autobiographical memories are needed to build a notion of personhood and identity, and our minds work to create a consistent set of identity "records," scaffolding the formation of identity that evolves over the years. The development of an autobiographical self is partly organized under genomic and biological control, and part of it is regulated by the environment—ranging from models of individual behavior to cultural rites. Remembering is vital to our well-being, because without autobiographical memories we would have no sense of past or future, and we would lack any sense of continuity. Our image of who we are, mentally and physically, is based on long-term remembrance of facts, emotions, and experiences; that self-image is never stable but is subject to constant remodeling because our perceptions of who we are change along with our projections and desires of who we want to be. As cognitive scientists argue, the key aspect of self-growth is to balance lived past with anticipated future.[2] Without the capability to form autobiographical memories—a defect that could happen as a result of partial brain damage—we are basically unable to create a sense of continuity in our personhood.

Grounded in the discourses of behavioral or social psychology, memory is also central to constructing a sense of a continuity between our selves and others. American psychologist Susan Bluck contends that autobiographical memory has three main functions: to preserve a sense of being a coherent person over time, to strengthen social bonds by sharing personal memories, and to use past experience to construct models to understand inner worlds of self and others.[3] Reminiscence allows people to reconstruct their lives through the looking glass of the present, and "cognitive editing" basically helps to bring one's present views into accord with the past. Of these three functions—self-continuity, communicative function, and directive function—Bluck regards the second as the most important one: people share individual experiences to make conversation more truthful, to elicit emphatic responses, or to develop intimacy and social bonds. In autobiographical memory, the self meets the social, as personal memories are often articulated by communicating them to others.

Expanding and refining Bluck's definition of autobiographical memory, psychologist Katherine Nelson identifies a cultural notion of self, in addition to the cognitive, social, and other levels of self-understanding psychologists have long recognized.[4] A cultural sense of self emerges around five to seven years of age, a developmental stage where children start to "make contrasts between the ideal self portrayed by the culture and the actual self as understood."[5] A child's autobiographical memory evolves as a

culturally framed consciousness, where personal narratives constantly intermingle with other stories: "Personal memories, which had been encapsu-
lated within the individual, become transformed through verbal narratives
into cultural memory, incorporating a cultural belief system."[6] A culturally
framed autobiographical memory integrates the sociocultural with the personal, and the self that emerges from this process is explicitly and implicitly
shaped by its environment's norms and values. As Nelson remarks, the narratives that confront children—fairy tales told by parents and teachers, or
stories they watch on television—are an important factor in their development. Children test their sense of self against the communal narratives they
are exposed to, either through verbal reports or via television or video. Even
though some cognitive and developmental studies on autobiographical
memory touch upon the important intersection of individual psychology
and socializing culture, few psychologists specify the role of culture in relation to memory. Wang and Brockmeier eminently expound on the interplay
between memory, self, and culture, arguing that autobiographical remembering manifests itself "through narrative forms and models that are culturally shaped and, in turn, shape the remembering culturally."[7] Even if (social)
psychologists acknowledge the dynamic relationship between memory and
self to be integrated in the larger fabric of a culture, and even if they affirm
that conceptions of self are inscribed in various material and symbolic
ways, the role (media) objects play in the process of remembering remains
largely unexamined. Understandably but regrettably, psychologists seem to
think those questions are the proper domain of anthropologists or media
scholars.

And yet, opening up sociopsychological perspectives on autobiographical memory to insights in cultural theory and media studies may
turn out to be mutually beneficial. Let me elucidate this by elaborating a
simple domestic scene from everyday life. A fifteen-month-old toddler attempts to stand on his own two feet and take his first cautious steps. His
parents are thrilled, and they converse about their relief over this happening. The delighted father brings out his video camera to capture the toddler's effort on tape; that same evening, the proud mother verbally reports
the first-step achievement to the grandparents. Snapshots of the child's developmental milestone, complemented by a few lines of explanation, sup-
plement the latest update on the family's website. The parents mark the event
through various activities: telling stories, taking pictures, and composing an

account help to interpret the event and communicate its significance to others. They concurrently produce material artifacts that may assist them—and their offspring—to recall the experience at a later moment in time, perhaps in different circumstances or contexts.

The autobiographical memory at work in this instance consists of several stages and layers—aspects that can be accentuated or eclipsed in consonance with respective academic interests. Psychologists center on how the parents interpret, communicate, and later recall baby's first steps. Mental frames and cognitive schemes help parents evaluate the event: they compare their own baby's achievement to infants' development in general. The average baby starts walking at twelve months, but this one is slower. Parents relate their experience in a narrative framework that places the event in the spectrum of their own lives and that of others. (How old was I when I started to walk? How old was the baby's sister? How slow or fast do babies in this family start walking?) Sharing their oral report with grandparents helps parents determine the significance of what happened, but it also sets the stage for later reminiscence: interpretation and narration form the mental frames by which the experience can be retrieved from memory at a later stage. Memory work thus involves a complex set of recursive activities that shape our inner worlds, reconciling past and present, allowing us to make sense of the world around us, and constructing an idea of continuity between self and others—the three functions Bluck describes, as noted earlier.

Cultural theorists considering this scene may shift the center of gravity and emphasize the way in which the parents record, share, and later reminisce about baby's first steps using various media. Recording the event through video, pictures, or a written account enhances its actual experience. Memory work involves the production of objects—in this case snapshots and video footage—with a double purpose: to document and communicate what happened. These items also portend future recall: for the parents to remind them of this occasion and for the baby to form a picture of what life looked like before his ability to register memories in the mind's eye. Later interpretations invariably revise the meaning of memories, regardless of the presence of hard evidence in the form of pictures or videos. In hindsight, baby's-first-step video may be viewed as an early sign of his lazy character, but it may also provide evidence of an emerging disability that went unnoticed at the time of recording.

Evidently, the same scene gives rise to two sets of inquiries into memory formation, each highlighting different aspects, and yet, the personal and cultural can hardly be disentangled because there is a constant productive tension between our (personal) inclinations to stake out certain events and the (social) frameworks through which we do so—between the (individual) activities of remembering and the (cultural) products of autobiographical recall. Acts and products of memory are far from arbitrary. In Western culture, filming and photographing baby's first steps are considered common ritualized attempts to freeze and store a milestone in a human being's development; hence, the decision of these parents to catch the event on film and arrest the moment in photographs is in tune with prevailing norms—norms that, naturally, change with every new generation and also vary culturally. Western European and American practices of remembering and recording significantly diverge from Asian or African mores in this area, due to diverging cultural norms and social relationships.[8] In general, personal memory stems from the altercation of individual acts and cultural norms—a tension we can trace in both the activity of remembering and in the object of memory.

Therefore, I want to define "personal cultural memory" as *the acts and products of remembering in which individuals engage to make sense of their lives in relation to the lives of others and to their surroundings, situating themselves in time and place.*[9] According to my definition, "personal" and "cultural" are the threads that bind memory's texture: they can be distinguished, but they never can be separated. We usually mark events because their significance is already ingrained in our conscious: first steps are an important happening in a child's life, just as birthdays and first school days are. The decision to record such events is already, to a large extent, stipulated by conventions prescribing which occurrences are symbolic or ritual highlights and thus worth flagging. Some events, such as conflicts or depressions, may seem unsuitable for video recording, but they may instead be amenable subjects for diary entries. Other events, such as household routines or intense emotions, are perhaps too dull or too poignant for any kind of inscription, yet that does not mean they cannot be recalled—most of our life's experiences, after all, go undocumented, and often deliberately so. Parents who decide *not* to take out their video camera may do so because they prefer to enjoy and remember the first-step experience without the camera's intervention. At various moments, people decide what to record or what to remember without records, often being unaware of the cultural frameworks that inform their intentions and prefigure their decisions. These frameworks, in other words, already inherently shape the functions of self-continuity, communication, and self-direction that memory work entails. Personal cultural memory entwines individual choice with common habits and cultural conventions, jointly defining the norms of what should be remembered.

What holds true for acts of memory also pertains to its ensuing products, particularly those created through media. Products of memory, whether they are family photographs, diaries, home videos, or scrapbooks, are rarely the result of a simple desire to produce a mnemonic aid or capture a moment for future recall. Instead, we may discern different intentions in the creation of memory products: we can take a picture just for the sake of photographing or to later share the photographed moment with friends. While taking a picture, we may yet be unaware of its future material form or use. However, any picture—or, for that matter, any diary entry or video take—even if ordained to end up in a specific format, may materialize in an unintended or unforeseen arrangement. In spite of the indeterminacy of a memory object's final reification—and this may sound paradoxical—familiar cultural formats always inherently frame or even generate their production. A range of cultural forms, such as diaries, personal photographs, and so on, configures people's choices of what they capture and how they capture it. For instance, family albums funnel our memories into particular venues; a rather extreme example may be the preformatted baby's first-year book, in which developmental signposts—from prenatal ultrasounds to first steps–pictures—are prescribed by its layout. These normative discursive strategies either explicitly or implicitly structure our agencies; I return to this issue in the next chapter, when discussing the meaning of digital technologies as memory tools, but suffice it to say here that existing models often direct our discursive means for communicating and remembering.

Therefore, it is a fallacy to think of memory products as purely constraining or conformist. They do not only enable structured expression but also invite subversion or parody, alternative or unconventional enunciations. Products of memory are first and foremost creative products, the provisional outcomes of confrontations between individual lives and culture at large. When discussing family albums or diaries, I often encounter prejudiced assessments that characterize these genres as boring, predictable, or bourgeois. Yet on closer inspection, it is quite remarkable how many people gain creative energy out of shaping their own histories and subjectivities in response to existing cultural frameworks.[10] Admittedly, few people record family rows, and though teenagers shooting home videos of their fathers' most irritable habits may count as exceptions, they nevertheless illustrate my point that the very presence of cultural forms incites individual expressions. It may not be a coincidence that many successful commercial productions (feature movies, television series, or published autobiographies) expound on playful, expansive versions of personal memory accounts.[11] Conventional formats for individual cultural memory thus both constrain and unfetter people's proclivity to inscribe experiences.

The term "personal cultural memory" allows for a conceptualization of memory that includes dimensions of identity and relationship, time and materiality. Temporal and material aspects are extensively theorized in the next chapter; for now, I dwell a bit more on the relational nature of personal cultural memory. The term emphasizes that some aspects of memory need to be explained from processes at work in our society that we commonly label as culture—mores, practices, traditions, technologies, mechanics, and routines—whereas these same processes contribute to, and derive from, the formation of individual identities. Yet by advocating a definition of cultural memory that highlights the significance of personal collections, I do not mean to disavow the import of collective culture. Quite to the contrary, if we acknowledge that individual preferences are filtered through cultural conventions or social frameworks, we are obliged to further explore the intricate connection between the individual and collective in the construction of cultural memory.

Individual versus Collective Cultural Memory

Collective memory, like its autobiographical penchant, is commonly referred to as something we have or lack: it is about our ability to build up a communal reservoir of relevant stories about our past and future, or about the human proclivity to forget things—such as amnesia of collective traumas or shameful episodes in our history. For the purpose of this book, I prefer the notion of cultural memory over collective memory because

I am less concerned with what these reservoirs do or do not consist of; instead, this book concentrates on how memory works in constructing a sense of individual identity and collectivity at the same time. To set up this claim, I first need to sketch how prevailing notions of collective memory have structured academic thinking, most notably in sociological and historical accounts.

Just as individual or autobiographical memory is almost automatically associated with theory formation in the area of psychology, collective memory, since the early twentieth century, has been the privileged domain of sociologists, historians, and cultural theorists. Originated in late nineteenth-century French and German sociology, the concept of collective memory was most prominently theorized by Maurice Halbwachs, a critical student of both Henri Bergson and Emile Durkheim. In *Les cadres sociaux de la memoire*, first published in 1925, Halbwachs sketches the partially overlapping *cadres* (spheres) of individual and larger communities, such as family, community, and nation. In contending that memory needs social frames, he distances himself from more physiological approaches to memory, particularly those insisting on the isolated enframing capacity of the human mind. Far from being a cognitive trajectory activated by internal or external stimuli, human memory "needs constant feeding from collective sources, just as collective memories are always sustained by social and moral props."[12] Halbwachs thus emphasizes the recursive nature of individual and collective memory, one always inhabiting the other. Collectivity, he claims, arises in the variable contexts of groups who share an orientation in time and space. Our memories organize themselves according to our actual or perceived participation in a (temporal) collectivity—a group vacation, a school class, a family, a generation—and recall tends to lean on a sense of belonging or sharing rather than on a relocation in real time or space. We may remember events chronologically or spatially, but quite often we remember in terms of connectivity. As social creatures, humans experience events in relation to others, whether or not these communal events affect them personally.

One of Halbwachs's important observations is that collective memory is never the plain sum of individual remembrances: every personal memory is cemented in an idiosyncratic perspective, but these perspectives never culminate into a singular collective view. The memories of both parent and child participating in the same event are not necessarily the same or even complementary: each partaker may retain vastly different interpretations of the occurrence. Yet even if their accounts are antithetical because of the different (social) positions of each member, they still "share" the memory of a communal event. Collectivity not only evolves around events or shared experience; it can also advance from objects or environments—anything from buildings to landscapes—through which people feel connected spatially. Halbwachs specifically draws attention to auditory expressions, such as music, voices, and sounds, to which people are exposed from an early age and that later serve as triggers for collective recall.[13] Each memory derived from these common resources can be distinctly different, and individual memories never add up to a collective reservoir.

Ever since Halbwachs coined the concept of collective memory, it has prominently figured in the accounts of historians, where it was also renamed "social" or "public" memory.[14] The historical meaning of "collective," however, differs from its sociological counterpart. In a sociological sense, "collective memory" means that people must feel they were somehow part of a communal past, experiencing a connection between what happened in general and how they were involved as individuals.[15] Adjusted to historiographical explanation, "social memory" constitutes the interface between individual and collective ordering of the past. Some historians have chosen collective memory as a central ordering concept for their interpretation of how history can be written. David Gross, who appropriates Halbwachs's term for the purpose of historiography, views (collective) memory as a prism for historical reconstruction: his main thesis concerns the value societies have placed in either remembering or forgetting as a basic life-orientation, and from this point of entry he reinterprets history from antiquity to late modernity.[16] Gross agrees that memory is a complicated encoding process and that memories are preserved through elaborate schemata and shaped by shifting forms, scripts, and social circumstances.[17] To properly understand their own existence in the grand scheme of historical events, people continuously sharpen their own remembered experience and the testimonies of others against available public versions—official documents, exhibits, text books, and so forth. Especially since WWII, historians are increasingly intrigued by the way in which personal accounts, or "small histories," reflect and refine the complexities of grand historical narratives. So-called ego-documents are now welcomed by official archives, museums, and other public "memory institutions."[18]

The recent institutionalization of personal memory items can be seen as a corollary to historians' designation of a "new" collective memory. And yet, the elevation of personal memory objects to the status of collective history's ingredients paradoxically underscores their distinct hierarchy. A case in point is the inclusion of numerous individual testimonies in public representations of the Holocaust. Especially in the past two decades, the collective remembrance of the genocide, after a period of relative suppression, has exploded into a plethora of forms: exhibitions, monuments, films, audio-visual testimonies, books, museums, and so forth. Taking the Holocaust exhibition at the Imperial War Museum in England as an example, historian Andrew Hoskins explains that its "mixing artifactual representations . . . with audio-visual mediation of individuals' memories of their experience of the Holocaust in the form of testimony of survivors" is a relatively recent phenomenon in public exhibitions.[19] Apart from the typically mediated nature of these testimonies—a pivotal aspect of Hoskins's characterization of "new" collective memory to which I turn in the next section—the relationship of innumerable individual accounts to collectivity seems self-evident and unproblematic. The aspiration to save all remaining individual testimonies of survivors to form a grand narrative of the Holocaust implicitly bolsters quite a few megaprojects such as Steven Spielberg's Shoah Foundation.[20]

However, as Halbwachs already observed, no collective experience—and certainly not one of this magnitude—can ever be represented in a singular collective memory. The inclusion in our public memory sites of many individual testimonies, each presenting a unique prism through which to make sense of historical events, will never add up to an overall collective view of the Holocaust. Disputing the view of some of his colleagues, American cultural historian Andreas Huyssen argues that the plethora of personal memories of the Holocaust may obscure rather than strengthen the notion of collective memory: "The problem for Holocaust memory in the 1980s and 1990s is not forgetting, but rather the ubiquitousness, even the excess of Holocaust imagery in our culture."[21] Huyssen questions the idea that individual memory representations serve as building blocks for, or form particular versions of, collective memory, because such a premise ignores the always inherent creative tension between individual and collective.[22]

Although the foregrounding of individual testimonies has undoubtedly helped popularize important takes on communal history, the assumed self-evident relationship between individual and collective memory is indeed problematic. Remarkable in both Halbwachs's sociological discourse as well as in Gross's historiographical account is the virtual absence of the term "culture." As an explanatory concept, cultural memory inherently accounts for the mutuality of individual and collective. Culture, like memory, is less interesting as something we have—hold or discount—than as something we create and through which we shape our personal and collective selves.[23] Like Halbwachs, I see the conjunction of individual and collective memory as dialectic, yet in emphasizing cultural memory, I stress the recursive dynamic of this ongoing interconnection beyond the level of cognition or sociality. Culture is more than the encounter of individuals with mental structures and social schemata, as Gross suggests; discursive and material artifacts, technologies, and practices are equally infested with culture, thus forming the interface between self and society.

Cultural memory is a guiding concept in the work of German historians Jan Assmann and Aleida Assmann. Building on Halbwachs's sociological theory, Jan Assmann defines cultural memory as "a collective concept for all knowledge that directs behavior and experience in the interactive framework of a society and one that obtains through generations in repeated societal practice and initiation."[24] Aleida Assmann expounds on this definition by sketching cultural memory as one end of a complex structure that also involves individual, social, and political memory—going from a purely private level to the institutionalized and ritualized level of remembrance.[25] She petitions a seamless transformation from individual to cultural memory, the result of which is never a fixed reservoir but always a relational vector that connects self to others, private to public, and individual to collective.[26] Unlike other historians, Assmann stresses the importance of memory objects' materiality in texts and images; the sum of individual objects of memory never add up to one unified "collective" memory—in fact, Assmann is very suspicious of this term—but the objects are unique anchors of remembering processes through which self and others become connected.

My own concept of cultural memory shows clear affinity with Aleida Assmann's dynamic definition. Perhaps more specifically, I prefer to think of cultural memory as an act of negotiation or struggle to define individuality and collectivity. Closely entwined with these two notions are the spheres of private and public; memory is as much about the privacy to

inscribe memories for oneself and the desire to share them only with desig-
nated recipients as it is about publicness, or the inclination to share experiences with a number of unknown viewers or readers. There is not, nor has
there ever been, a sharp distinction between private and public, but every
act of memory involves a negotiation of these spheres' boundaries. The intention to inscribe or recall a memory exclusively for private use may change
over time, as personal memory may acquire a larger significance against a
background of evolving social mores or personal growth. Control over one's
memory may also change in the course of time; one may lose command
over either one's mental capacities for remembering, as a result of disease or
death, or one may lose ownership over material inscriptions of former experiences, whether voluntary or involuntary. Intentions and control change
along with our revisions of memories in the passage of time, and revisions,
in turn, reset the boundaries for what counts as public or private. Those
boundaries are concurrently the outcome and stakes in the act of cultural
memory.

Let me illustrate this specified concept of cultural memory using an
example—an example I further elaborate in Chapter 3. Anne Frank's diary
is most commonly typified as the poignant personal lens through which we
experience a collective memory of the Holocaust. From my perspective,
though, Frank's diary stands for a continuous and ongoing struggle between individual and collective acts of memory. Defined as personal
cultural memory, it signifies a Dutch teenager's choice to narrate her experience in a cultural form—a handwritten daily account, trusted to a notebook, that she later revised; Otto Frank's decision to publish selected parts
of his daughter's journal turned the diary into a public, collective item.
Anne's aspiration to become a novelist as well as her father's judgment to
censor the first editions should be understood in the context of the larger
cultural arena in which these mandates were negotiated. Naturally, the ensuing Anne Frank industry—the museum, the objects in the museum, the
play, the movies, television series—are part of the (collective) cultural act
of remembrance, but they are also products of the memory industry.[27] All
past, present, and future choices made in the service of inscribing and preserving Anne Frank's legacy are in fact collusions of individuality and collectivity. Memory filtered through the prism of culture acknowledges the
idea that individual expressions are articulated as part of, as much as in
spite of, larger collectivities; individuality can be traced in every negotiation

of collectivity—past and present—as it is always a response to all previous representations.

In the disciplines of the humanities, cultural memory seems to automatically refer to collective remembering, whether or not as a subset of history, just as autobiographical memory appears to be the realm of the individual psyche, indeed operating as part of a social collective but always subordinate to it. By default, the term "cultural" has come to reflect collectiveness, whereas the term "autobiographical" connotes individuality. My argument that the term "cultural" inherently relates individual and shared memory should in fact render the preceding qualifiers "personal" or "collective" to cultural memory redundant. However, use of the modifier "personal" indexes the impossibility of insulating the individual from culture at large. Mutatis mutandis, when speaking of collective cultural memory, the term inherently accounts for those individuals creating collectivity and through whose experiences and acts culture is constituted. Even if my choice of terminology seems cumbersome, it is prompted as much by the genealogy of disciplinary appropriation as by a desire to stress the relational nature of these terms: cultural memory can only be properly understood as a result of individual's and others' mutual, interdependent relationship.

As much as I appreciate Aleida Assmann's conceptual clarity, something is missing from her model of cultural memory that appears to be highly relevant to further translation of this concept into a usable model of analysis. Although she stresses the interference of mental and cultural frameworks in her theory, she clearly does not know how to account for the role of media and media tools in the formation of cultural memory. Like other historians, she refers to media as templates or repositories molding our experiences, and she considers media to be problematic in the way they profoundly affect memory discourse. As noted earlier, psychologists also allude to media (or media frames) as collective narrative forms affecting the individual psyche, but few proceed to include this alleged influence in their theoretical models. It is peculiar to notice how memory scholars recognize media to play a considerable role in the construction and retention of experience; and yet, media and memory are often considered two distinct—sometimes even antagonistic—domains. Therefore, I now shift attention to the mediation of memory to find out how we can render media an integral element of our new analytical tool.

The Mediation of Memory

In recent years, cultural theorists have observed an irreversible trend toward what is generally called the "mediation of memory," the idea that media and memory increasingly coil beyond distinction.[28] Although the mediation of memory is an important concept, it also carries some internal contradictions. Before proposing my modified concept, I first need to argue what exactly is wrong with the models that pass in review. Many theories acknowledge the intimate relationship between memory and media, but that union is often contingent on a set of fallacious binary oppositions. First, there is the tendency to discern memory as an internal, physiological human capacity and media as external tools to which part of this human capability is outsourced. Adjunct to this distinction is the implicit or explicit separation of real (corporeal) and artificial (technological) memory. Third, media are qualified either in terms of their private use or of their public deployment, as mediators of respectively personal or collective memory.

Over the years, both negative and positive appreciations of media and memory's alliance reveal such binary thinking. From the days of Plato, who viewed the invention of writing and script as a degeneration of pure memory (meaning: untainted by technology), every new means of outsourcing our physical capacity to remember has generated resentment.[29] Most scholars acknowledge the continuation of memory's "technologization"— a term powerfully argued by Walter Ong—from manual and mechanical means of inscription, such as pencils and printing presses, all the way to modern electronic and digital tools; and yet, they often only refer to the more recent stages as "technologically mediated."[30] With the advent of photography, and later film and television, writing tacitly transformed into an interior means of consciousness and remembrance, whereupon electronic forms of media received the artificiality label. Due to the rise of electronic images, often held responsible for the decline of the printed word, writing gained status as a more authentic container of past recollection—an irony likely to recur with each new generation of technologies.[31] Besides generating resentment, the emergence of electronic (external) memory has also been applauded, an appreciation that often has been based on the very same bifurcated models. In the influential theories of Marshall McLuhan, electronic media, as "extensions of men," signaled the unprecedented enhancement of human perceptual capacities: photography and television were augmentations of the eye whereas audio technologies and radio extended the ear's function.[32]

Similar dualities can be traced in more recent debates, particularly those discussing how mass media infiltrate collective memory. Adding to the disjunction is the tendency to define memory in ambiguous terms of media: either as tools for inscribing the past or as an archival resource. For instance, when French historian Pierre Nora laments the enormous weight of media versions of the past on our historiography, he basically regards collective memory as a giant storehouse, archive, or library.[33] Contrasting this conceptualization is Jacques Le Goff's concern that media representations form a filter through which the past is artificially ordered and edited—manufactured rather than registered.[34] At once a means of inscription and an external repository, media are seen as apparatuses for production and storage, modeled after the mind's alleged capacity to register and hold experiences or impressions. Visions of printed and electronic media as replacements of human memory notably echo in phrases like this one from the late British historian Raphael Samuel: "Memory-keeping is a function increasingly assigned to the electronic media, while a new awareness of the artifice of representation casts a cloud of suspicion over the documentation of the past."[35] Even if unarticulated, pervasive dichotomies inform scholarly assessments of the media's role in the process of remembrance. On the one hand, media are considered aids to human memory, but on the other hand, they are conceived as a threat to the purity of remembrance. As an artificial prosthesis, they can free the brain of unnecessary burdens and allow more space for creative activity; as a replacement, they can corrupt memory. Media are thus paradoxically defined as invaluable yet insidious tools for memory—a paradox that may arise from the tendency to simultaneously insist on the division between memory and media and yet conflate their meanings.[36]

Media and memory, however, are not separate entities—the first enhancing, corrupting, extending, replacing the second—but media invariably and inherently shape our personal memories, warranting the term "mediation." Psychologists point at the inextricable interconnections between acts of remembrance and the specific mediated objects through which these acts materialize. As Annette Kuhn claims, photographic images, "far from being transparent renderings of a pre-existing reality, embody coded references to, and even help construct, realities."[37] Mediated memory objects never represent a fixed moment; they serve to fix temporal notions and relations between past and present. Steven Rose, a British scientist who discusses the physiological complexity of human memory and consciousness at length, makes the case that mnemonic aids, such as photos or videos, are confounded with our individual memories to such extent that we can hardly distinguish between the two.[38] And anthropologist Richard Chalfen, in his study of how home media help communicate individuals' perceptions of self to others, argues that our "snapshots are us."[39] Personal cultural memory seems to be predatory on media technologies' and media objects' shaping power.

A similar notion of mediation can be found among historians discussing the infiltration of mass media in collective memory. British sociologist John Urry explains in his essay "How Societies Remember the Past" that electronic media intrinsically change the way we create images of the past in the present.[40] Mass media, according to historian of popular culture George Lipsitz, embody some of our deepest hopes and engage some of our most profound sympathies; films, records, or other cultural expressions constitute "a repository of collective memory that places immediate experience in the context of change over time."[41] Media like television and, more recently, computers are devices that produce, store, and reshape earlier versions of history. With the recent explosion of electronic and digital devices, the tools for memory have shifted in nature and function—a subject I return to in the next chapter.

We can witness this contrived interlocking most poignantly at the metaphorical level; the term "mediation of memory" refers equally to our understanding of media in terms of memory (as illustrated by historians' accounts above) as well as to our comprehension of physiological memory in terms of media, evidenced by the many metaphors explaining certain features of human memory. Ever since the invention of writing tools, but most noticeably since the emergence of photography in the nineteenth century, the human capacity to remember has been indexed in daily language by referring to technical tools for reproduction. Dutch psychologist Douwe Draaisma, who extensively researched the historical evolution of memory metaphors, notes that media are a special conceptual category for envisioning memory's mechanics.[42] For instance, the term "flashbulb memory"—the proclivity to remember an impacting moment in full detail, including the circumstances, time, and place in which the message was received—derived its signifier from the realm of photography. In the twentieth century, the terminology of film and video started to invade the discourse of memory and memory research: life is said to be replayed like a film in the seconds before death, and psychologists have extensively examined the phenomenon known as "deathbed flash." By the same token, we are now firmly grilled by the media's convention to visualize a character's recall of past experience as a slow-motion replay or flashback. Metaphors are not simply means of expression, but are conceptual images that structure and give meaning to our lives.[43] As media have become our foremost tools for memory, metaphorical reciprocity signals their constitutive quality.

However compelling and valid, most mediation of memory theories still hinge on a few premises that funnel the scope of this concept, restricting its explanatory range. First, an exclusive focus on *either* home media *or* mass media often presume a symbiotic union of home media with individual memory and of mass media with its collective counterpart; such rigid distinctions hamper a fuller understanding of how individual and collective memory are shaped in conjunction, with media serving as relational means connecting self to others. Second, even though most theories acknowledge the convergence of memory and media, the "mediation" concept frequently favors a single vector: media shape our memories, but we seldom find testimony of media being shaped by memory, indicating an implicit hierarchy. Let me address each of these conceptual deficiencies in more detail.

It is practical to assume that personal cultural memory is generated by what we call home media (family photography, home videos, tape recorders) whereas collective cultural memory is produced by mass media (television, music records, professional photography), implying that the first type of media is confined to the private sphere, whereas the latter pertains to the public realm. But that simple division, even if functional, is also conceptually flawed: it obscures the fact that people derive their autobiographical memories from both personal and collective media sources. Media sociologist John Thompson, who highlights the role of individual agency in media reception, explains the hermeneutic nature of this relationship.[44] He argues that "lived experience," in our contemporary culture, is interlaced with "mediated experience"; mediation, then, comprises not only the media tools we wield in the private sphere but also the active choices of individuals to incorporate parts of culture into their lives. Experience is neither completely lived nor entirely mediated, as the encounter between the two is a continuously evolving life-project to define the self in a larger cultural context. What makes mediated experience today differ from lived experience two hundred years ago is the fact that individuals need no longer share a common locale to pursue commonality; the growing availability of mediated experience creates "new opportunities, new options, new arenas for self-experimentation."[45] If we accept a preliminary distinction between home and mass media, we not only fail to account for media shaping our sense of individuality *and* collectivity in conjunction, but we equally obscure how individuals actively contribute to the collective media that shape their individuality.

A strict delineation of home and mass media has also become impracticable because each type informs the other. For instance, *America's Funniest Home Videos* (*AFHV*), a program format that has been successfully franchised to television stations around the globe for over two decades, is made up entirely of home media footage, woven into a glitzy commercial production; by the same token, many home video enthusiasts have taken *AFV*'s narrative structure (or its announced themes) as directive models to film their cute children and pets. Amateur videos of the tsunami hitting the shores of Thailand, Sri Lanka, and Indonesia in December 2004 made the news worldwide; the power of amateur footage probably motivates individuals who buy and carry a camera to create news and offer it to television stations. There are many examples of documentaries based largely on compilations of amateur or home movies, and this trend is likely to increase with the growing availability of affordable digital cameras, computers, and editing equipment. In the next chapter, I further elaborate on the meaning of media technologies (digital and analog) as tools that mediate between personal and collective cultural memory.

The second conceptual deficiency in the mediation of memory concerns the implied hierarchy between external and internal memory, or in plain terms, between technology and the human mind. Understandably, our metaphors often explain the invisible in terms of the visible and knowable; that is why the mysteries of mental processing are often elucidated in terms of media technology—technology that is at once transparent and mechanically predictable. But how about turning the vector back on its arrow: could our development and use of various media technologies be informed by the perceptual mechanisms, the sensory motor actions that underlie memory formation?[46] An intricate aspect of remembering is that mental perceptions (ideas, impressions, insights, feelings) manifest themselves through specific sensory modes (sounds, images, smells). The media we have invented and nursed to maturity over the years incorporate a similar tendency to capture ideas or experiences in sensory inscriptions, such as spoken or written words, still or moving images, recorded sounds or music. One memory rarely encompasses all sensory modes, because we tend to remember by selecting particular ones. For instance, we may recall a mood, locale, or era through a particular smell (such as the smell of apple pie in the oven triggering the image of your mother's kitchen on Saturday afternoons), or we may remember a person by his nasal voice or her twinkling eyes.[47] The same holds true for memories captured through media technologies: rather than wanting exhaustive recordings, we commonly select a specific evocative frame in which to store a particular aspect of memory—a still photograph to store visual aspects, a diary entry to retain interpretative details and subjective reflection, or a video to capture the movement of baby's first steps. We have a gamut of preferred sensory and medial modes at our disposal to inscribe specific memories, but the intriguing question is: do available media technologies dictate which sensory aspects of an event we inscribe in our memory, or do sensory perceptions dictate which medium we choose to record the experience?

Although most of our media technologies privilege a particular sense (e.g., photography singles out sight, tape recorders sound) that does not mean there is a one-to-one relationship between specific sensorial aspects of memory and the preferred instrument of recording. On the contrary, a still picture may invoke the sound of a child's laughter long after the child has grown into adulthood. Nevertheless, instruments of memory inscription, privileging particular sensorial perceptions over others, always to some extent define the shape of our future recall. Most people have unconscious preferences for a particular mode of inscription; for instance, they favor moving images over still pictures, or oral accounts over written. Although part of that propensity is undoubtedly rooted in individual mindsets, another part is inevitably defined by the cultural apparatus available and socially accepted at that time. But this apparatus is far from static: each time frame redefines the mutual shaping of mind and technology as one is always implied in the other. The brain steers and stimulates the camera as much as the camera stimulates the brain. Mediated experience, as film scholar Thomas Elsaesser remarks, is subjected to a generalized cultural condition:

In our mobility, we are "tour"ists of life: we use the camcorder in our hand or often merely in our head, to reassure ourselves that this is "me, now, here." Our experience of the present is always already (media) memory, and this memory represents the recaptured attempt at self-presence: possessing the experience in order to possess the memory, in order to possess the self.[48]

In other words, memory is not mediated by media, but media and memory transform each other. In the next chapter, I further elaborate on how changing (digital) technologies and objects embody changing memories.

A revised concept of memory's mediation thus needs to avoid the pitfalls of fallacious binary and hierarchical structures. Media are not confined to private or public areas, and neither do they store or distort the past in relation to the present or future. Like memories, media's dynamic nature constitutes constantly evolving relations between self and others, private and public, past and future. The term "mediation of memory," if we attend to its conceptual flaws, may be rearticulated as "mediated memories," a concept that may add to a better understanding of the mutual shaping of memory and media.

Mediated Memories as Objects of Cultural Analysis

Mediated memories are the activities and objects we produce and appropriate by means of media technologies, for creating and re-creating a sense of past, present, and future of ourselves in relation to others. Mediated memory objects and acts are crucial sites for negotiating the relationship between self and culture at large, between what counts as private and what as public, and how individuality relates to collectivity. As stilled moments in the present, mediated memories reflect and construct intersections between past and future—remembering and projecting lived experience. Mediated memories are not static objects or repositories but dynamic relationships that evolve along two axes: a horizontal axis expressing relational identity and a vertical axis articulating time. Neither axis is immobile: memories

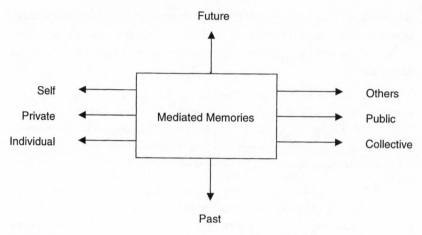

FIGURE I

move back and forth between the personal and collective, and they travel up and down between past and future. The commingling of both axes signals memories should be understood as processes—mediating self to culture to past to future. The dynamic nature of this analytical model shows in Figure 1.

Mediated memories refers both to acts of memory (construing a relational identity etched in dimensions of time) and to memory products (personal memory objects as sites where individual minds and collective cultures meet). The term is neither a displacement of a psychological definition of (personal) cultural memory, nor is it a dislodgment of the historian's notion of (collective) cultural memory. Instead, it offers a tool for analyzing complex cases of memory formation and transformation, taking personal shoebox items as culturally relevant objects.

Taken at face value, my mediated memories concept appears quite similar to the term "tangled memories" coined by American media theorist Marita Sturken. Sturken deploys a concept of cultural memory that comes close to Aleida Assmann's, but unlike the German historian, she includes media (objects) in her theoretical model, even turning them into a central focus of analysis.[49] Sturken explains how items of popular culture—from the Vietnam Memorial to the AIDS Memorial Quilt—are compiled from personal acts of commemoration out of which arises a collective statement about a shared trauma. Media play a principal role in this process, both as personal instruments (e.g., private photos or videos) and as mass conveyors of social narratives (e.g., news coverage or documentaries). Individuals root their personal memories in cultural objects and activities that at once give meaning to their own experience and attribute historical or political significance to the larger events of which they are part. Sturken correctly observes that personal memory, cultural memory, and history do not exist within neatly defined boundaries: "Memories and memory objects can move from one realm to another, shifting meaning and context. Thus personal memories can sometimes be subsumed into history and elements of cultural memory can exist in concert with historical narratives."[50] If we compare Sturken's concept of tangled memories to my concept of mediated memories figured in the diagram, they differ on two accounts. As figure 1 shows, it is possible to advance mediated memories from both the right and the left angle. Sturken approaches manifestations of cultural memory from the (right) angle of collectivity. What is more, she explicitly excludes "shoe-box contents" insofar as they remain private possessions, not having gained any cultural or political relevance in the public, collective realm.[51] Sturken's approach is similar to Alison Landsberg's who proposes the term "prosthetic memory" to argue that memories are not so much socially constructed or individual occurrences but emerge at the interface of individual and collective experience; she states that prosthetic memory is "less interested in large-scale social implications and dialectics than in the experiential quality of prosthetic memory and in the ramifications of these memories for individual subjectivity and political consciousness."[52]

Both Sturken and Landsberg approach cultural memory from the right angle of collectivity, showing how memories have the ability to alter a person's (political) outlook or action. In contrast, this book approaches memories from the opposite direction, privileging private memory objects, regardless of whether they have gained recognition in the public realm. By emphasizing the left angle, we acknowledge cognitive or psychosocial dimensions of remembering as complementary to historical, political or cultural dimensions of memory. My thesis that we cannot separate the individual psyche from culture is articulated in the relational (mediating) nature of the diagram, hence prompting the integration of psychological and cognitive perspectives into cultural theory.

There is one more aspect in my proposed concept that I wish to push to the fore: the term "mediated memories" is intended as a conceptual tool for the analysis of dynamic, *continuously changing* memory artifacts and items of mediated culture. People wield photo and video cameras, computers, pens and pencils, audio technologies and so on to record moments in which lived experience intermingles with mediated experience. During later stages of recall, they may alter the mediation of their records so as to relive, adjust, change, revise, or even erase previously inscribed moments as part of a continuous project of self-formation. Films or photos are not "memory"; they are mediated building blocks that we mold in the process of remembering. Concrete objects stand for relational acts of memory; collections of mediated objects, stored in shoeboxes, often become the material and symbolic connection between generations whose perception of family or self changes over time, partly due to larger social and cultural transformations, partly depending on the intergenerational continuity each family member brings into this heritage. Beyond immediate family circles, material inscriptions may become part of a more public project—for instance a documentary—and thus add to a shared collective remembrance. In any case, mediated memories never remain the same in the course of time but are constantly prone to the vagaries of time and changing relations between self and others.

As stated in the introduction to this chapter, a shoebox may contain both self-made media objects and pieces of mediated culture. We increasingly wield technologies to record selected pieces of culture, items considered worthwhile additions to our personal collections. In fact, every decision to buy a book or record, or to tape a television program, situates a person in his or her contemporary culture. Engaging in commercial transactions, like buying music CDs or television series in DVD boxes, differs from deploying the same equipment (tape and cassette recorders, CD players and recorders, VCRs, MP3 players, and computers) to create one's own content, but both acts are immanent to the construction of cultural memory.[53] By selecting, recording, rerecording, and sharing assorted items of mediated culture, people build up their personal collections as an evolving project of self-formation. Since the 1950s, we no longer need to derive our personal tastes or cultural preferences mainly from social circles close to us, because media have expanded the potential reservoir for cultural exchange to much larger, even global, proportions. Whether with friends or family or with complete strangers, exchanging self-recorded items is also a way of creating collectivity—the "distanced commonality" that Thompson considers a pivotal feature of mediated experience. As a result, our personal collections of taped music, videos, and so forth can be considered markers of cultural agency; it is through these creative rerecordings and recollections that cultural heritage becomes established.

Through the looking glass of cultural analysis, mediated memories magnify the intersections between personal and collective, past and future; as acts and objects of memory, they involve individuals carving out their places in history, defining personal remembrance in the face of larger cultural frameworks. Individuals make selections from a culture surrounding them, yet they concurrently shape that collectivity we call culture. Cultural memory, hence, is not an epistemological category—something we can have, lack, or lose—and neither is mediation the inevitable effacement, distortion, or enhancement of human memory. Rather, cultural memory can be viewed as a process and performance, the understanding of which is indispensable to the perennial human activity of building social systems for cultural connectivity. Mediated memories reflect this cultural process played out by various agents—individuals, technologies, conventions, institutions, and so forth—whose acts and products we should examine as confrontations between individuality and collectivity. Counteracting the overwhelming emphasis on collective cultural memory, I aim to restore attention to the mediated items in our shoeboxes as collections worthy of academic scrutiny. Personal memory can only exist in relation to collective memory: in order to remember ourselves, we have to constantly align and gauge the individual with the collective, but the sum of individual memories never equals collectivity. Moreover, I refute the presumption that shoeboxes are only interesting in hindsight, after history has decided whether they contain material worthwhile of illustrating particular strands of the grand narrative. Our private shoeboxes are interesting on their own devices, as stilled cultural acts and artifacts, teaching us about the ways we deploy media technologies to situate ourselves in contemporary and past cultures and how we store and reshape our images of self, family, and community in the course of living.

In the next chapter, I turn to the so far underexposed aspect of mediated memories' materiality and technology, further exploring the two-axes model in light of recent cultural and technological transformations.

Because we are currently in the midst of a transition from analog to digital media, we need to pay even more attention to mediated objects as sites of cultural contestation. In order to succeed, we may need to further entwine perspectives of the biological and cognitive sciences with the insights of social sciences and humanities.

2

Memory Matters in the Digital Age

In the movie *Eternal Sunshine of the Spotless Mind*, the company Lacuna Inc. advertises its method for focused memory removal with the following slogan: "Why remember a destructive love affair if you can erase it?"[1] When Joel Barish (Jim Carrey) incidentally finds out that his ex-girlfriend Clementine Kruczynski (Kate Winslet) has undergone the Lacuna procedure to wipe their bitterly ended relationship from her memory, he requests Dr. Howard Mierzwiak to perform the same procedure on his brain. Joel is instructed to go home and collect any objects or mementos that have any ties to Clementine ("photos, gifts, CDs you bought together, journal pages") and to bring them to the doctor's office. Upon his return, Lacuna-technician Stan wires Joel's brain to a computerized headset; the doctor holds up each separate object (drawings from his diary, a mug with Clementine's picture, etc.) and tells Joel to let each object trigger spontaneous memories. Stan subsequently tags each object-related memory and punches it into a computer, apparently recording Joel's mental associations on a digital map of Clementine. That same night, Stan and his assistant come to Joel's house, hook up their drug-induced sleeping client to a machine that looks like a hairdryer but generates images similar to those produced by functional magnetic resonance imaging (fMRI), and start the erasure process. As the Lacuna Inc. website explains: "The procedure works on a reverse timeline, which means it begins with the most recent memories and goes backward in time. This approach is designed to target the emotional core that every memory builds on. By eradicating the core, Dr. Mierzwiak is able to make the entire memory dissolve."[2] One by one, Joel's memories of Clementine are erased—a fairly automatic process that would have been finished by early morning if not for Joel's realization, halfway through the procedure, that he wants to keep the good memories of his love affair, so he actively starts to resist the erasure guys. Incapacitated by drugs, he embarks on a dreamlike, psychic journey with a remembered Clementine, creatively hiding her in unconscious, untargeted corners of his memory where she does not belong, in an attempt to escape the high-tech apparatus that is slowly stripping away Joel's recollection of his former girlfriend.

Michel Gondry's and Charlie Kaufman's fictional treatise of modern science's struggle to erase undesirable autobiographical memories raises important questions: First, what is the "matter" of personal memories? Memory is obviously embodied, but neurobiologists, cognitive philosophers, and cultural theorists hold different—even if complementary—views on what "substance" memories are made of. Scientific concepts of memory have evolved significantly in recent decades, and the movie actually reflects on some recent neurocognitive theories on memory formation and retrieval. Second, *Eternal Sunshine of the Spotless Mind* presents an ambiguous answer to the issue of where memory is located. On the one hand, personal memory is situated inside the brain—the deepest, most intimate physical space of the human body. On the other hand, personal memories seem to be located in the many objects Joel and Clementine (like most of us) create to serve as reminders of lived experiences. Most of these items are what, in the previous chapter, I dubbed "mediated memory objects," such as pictures, videos, recorded music, diaries, and so on; people have a vested interest in them because they come to serve as material triggers of personal memories. Mediated memory objects, however, are not simply prostheses of the mind, as the movie wants us to believe. Mediated memories, as I argue in this chapter, can be located neither strictly in the brain nor wholly outside in (material) culture but exist in both concurrently, for they are manifestations of a complex interaction between brain, material objects, and the cultural matrix from which they arise.

After exploring how mediated memories are concurrently embodied through the mind and brain, enabled by media technologies, and embedded in a cultural context, the question arises, what happens when memory production enters the digital era? What makes the movie *Eternal Sunshine* interesting in this respect is the conversion of Joel's painful memories into digital brain scans. The technology deployed by Lacuna's technicians transcribes memories triggered by material objects onto a digital map that looks like a series of brain scans. Like ordinary computer files, the Clementine files can be erased and thus be made to disappear from the place in Joel's brain where they are stored. The proposed translations from experiences into memory objects, back into actual memories, and then into information files—files that can subsequently be stored or deleted—propel sophisticated perspectives on the (de)materialization and (dis)embodiment of memory. In contrast, the movie's depiction of memory objects in the digital age is rather simplistic: photographs, scrapbooks, and cassette tapes still seem to dominate Joel's and Clementine's mutual recollections in a period when digital pictures, weblogs, compact disks and MP3 files are rapidly replacing their analog precursors. In light of recent revisionist theory on memory formation, the question arises how digital technologies may accommodate the "matter" of both mind and media. Even if memory in the digital age is just as embodied and mediated through artifacts as before, the very notions of embodiment and materiality need upgrading, in order to account for memory's morphing nature. Upon entering the digital era, the question of where mediated memories are located or produced—how they are embodied, enabled, and embedded—becomes even more poignant.

Embodied Memory: "Personal Memory Is in the Brain"

Ever since memory entered scientific discussions, it has been caught in the brain-mind dichotomy and appropriated by scholars from various disciplines. Whereas philosophers tend to confine acts of memory to the mind, (neuro) scientists concentrate on the brain as the locus of memory's origin. Until the early twentieth century, the location of memory was generally consigned to the mind, and the stuff that memories were thought to be made of—an indefinable, immaterial set of thoughts and mental productivity—was considered the province of philosophers. From John Sutton's rather impressive historiography of how philosophers from Augustine to Descartes and from Hume to Bergson have conceptualized memory, it transpires that former spatial concepts of thinking about memory have gradually given way to connectionist concepts.[3] Metaphors such as the library and the archive were commonly used to explain the retention of information or the preservation of experience in an enclosed space, from where it can be retrieved on command.[4] When trying to remember something, the mind, triggered by a material object or image, searches through the stacks from which stored and unchanged information can be retrieved and reread. Research paradigms based on these metaphors assumed memories to be static data from someone's past, and this assumption still often exists in popular representations of memory.

In his important work *Matter and Memory* (1896), the French philosopher Henri Bergson already refuted a one-to-one correspondence between physical stimuli and mental images to account for human consciousness, instead proposing a recursive relationship between material triggers and the images formed by our minds.[5] Bergson's view that memory is not exclusively a cognitive process but also an action-oriented response of a living subject to stimuli in his or her external environment prohibits the idea of a pure memory preceding its materialization in a mental image. According to Bergson, "to picture is not to remember," meaning that the present summons action whereas the past is essentially powerless; recollection images are never re-livings of past experiences, but they are actions of the contemporary brain through which past sensations are evoked and filtered. In chapter 3 of *Matter and Memory*, Bergson discusses the relationship between pure memory, memory image, and perception. In order to analyze memory, he states, we have to follow the movement of memory at work. In that movement, the present dictates memories of the past: memory always has one foot in the present and another one in the future. The brain does not store memories but re-creates the past each time it is invoked: "The bodily memory, made up of the sum of the sensori-motor systems organized by habit is a quasi-instantaneous memory to which the true memory of the past serves as a base."[6] In other words, rather than accepting the existence of a reservoir of pure memory from which the subject derives its remembrances, Bergson theorizes that the image invoked is a construction of the present subject. The brain is less a reservoir than a telephone system: its function is to (dis)connect the body, to put the body to action or make it move.

This shift toward a connectionist model of understanding the matter that memory is made of definitely transpires from recent research by scientists studying its neurological and genetic workings; they point at the brain as the nucleus of all our mental activity and consciousness.[7] Genes, neurons, and living cells all constitute the bodily apparatus needed to carry out mental functions, for instance cognitive tasks such as factual recall, or affective tasks such as emotions or feelings. In spite of putting the center for memory activity in the brain, scientists assert there is no such thing as a single location for memory. Even though some parts of the brain are specialized in specific memory tasks—such as the hippocampus for retaining short- and long-term memory, the amygdala for emotional learning—there is no single vector between one brain system and one type of memory. Autobiographical memory is usually associated with emotional matters that are in turn sheltered by the two amygdalae, yet this does not mean they are solely confined to this part of the brain. Instead, the establishment of memories depends on the working of the entire brain network, consisting in turn of several memory systems, including semantic and episodic memory, declarative or procedural memory.[8] The hunt for the location of memory, undertaken by scientists of various disciplines, has come up with a staggering distributed answer to that question, in fact defying the very possibility of pinning down one type of memory to a single place in the brain. Facilitated by neurological circuits, the brain sets the mind to work, stimulating a perception or a mode of thinking—a mental image, a feeling—that in turn affects our bodily state. The brain is thus the generator of reflexes, responses, drives, emotions and, ultimately, feelings; memory involves both (the perception of) a certain body state and a certain mind state.

In more recent philosophies of mind, connectionist metaphors tend to conceive of memory as a distributed agency that leaves traces of an ongoing process. Of all connectionist metaphors that philosophers and neuroscientists have introduced over the years, the networked computer is probably the most prominent one, but it is not necessarily the best one.[9] Perhaps the symphony orchestra is a more appropriate metaphor than the computer when it comes to explaining the function of memory and how the brain's matter is responsible for the personal memories it produces.[10] Like a performance of Mahler's Seventh Symphony by an orchestra requires a brass section, a string section, and a percussion section, memory is a function of the *brain* that manifests itself through the mind and directs our consciousness or conscious acts, such as self-reflection or autobiographical reminiscence; it is a consortium of concerted efforts resulting in a momentary performance. Each member of the orchestra plays his or her part, following the prescribed score as well as the conductor's instructions— their individual performances contributing to the overall sense of harmony. The composer's notational score may be adjusted under the influence of some single parts or as a result of the audience's interpretation or appreciation. Even the hardware of musical instruments may be tweaked to accommodate the performance; material changes in musical instruments inevitably result in subtle performative changes. And, as every music aficionado knows, a symphony's performance changes over time, as each performer tends to interpret the score as well as previous performances through a contemporary ear.

The extended symphony metaphor may also account for why memories change each time they are "performed" by the brain. From recent neuroscientific research, we learn that the brain stores emotional memories very differently from unemotional ones. Negative emotional memories are retained in much more detail than positive emotional memories. In the case of traumatic memories, they tend to be captured by two separate parts of the brain: the hippocampus, the normal seat of (cognitive) memory; and the amygdala, one of the brain's emotional centers. Hippocampal damage can affect one's capability to form long-term memories, but someone suffering from this condition may still be able to recall vague pleasant memories if the amygdala is left intact. Memories effectively are rewritten each time they are activated; instead of recalling a memory that has been stored some time ago, the brain is forging it all over again in a new associative context. Every memory, therefore, is a new memory because it is shaped (or reconsolidated) by the changes that have happened to our brain since the memory last occurred to us. Neuroscientists' findings are corroborated by clinical psychologists whose research demonstrates that memories of personal experience are never direct and unalterable copies of past experiences but are partially reconstructed; self and memory work in tandem to allow us the ability to use our own past as a present resource.[11]

In more than one respect, the movie *Eternal Sunshine* appears in sync with current neuroscientific research, as it demonstrates a nuanced understanding of how the brain forms memories. Scientist Steven Johnson states in his review of the film that whereas older movies like *Memento* still reflect the idea of memory as a kind of information retrieval system, the "emphasis on feeling over data processing puts *Eternal Sunshine* squarely in the mainstream of the brain sciences today."[12] In *Memento*, the main character, Leonard, suffers from complete amnesia after a major trauma; he fervently tries to reconstruct his past by taking snapshots, which he instantly annotates with words—sometimes tattooed on his body—in a desperate attempt to counter the constant loss of information about his own identity and past experiences from his brain.[13] Lost information or memories seem to be fixed in the past and are fixated in the annotated photographs Leonard keeps producing. In contrast, *Eternal Sunshine* reflects a more complex model of memory and how it is stored in different centers of the brain. Joel's memories partly consist of information that can be erased, yet their emotional core persists. Moreover, his memories are not fixed but morph into new ones. When Joel realizes he needs to stop the erasure procedure, he consciously manipulates the process by taking Clementine to memory spaces where she does not belong—kidnapping her away from the probing scanner, ushering her into scenes from his childhood that he remembers as being humiliating, painful, or very happy. These intense emotional memories are not so much reexperienced as they are rewritten through his recollection. Without the slightest science babble, the movie's assumptions on autobiographical memory are broadly compatible with the reconsolidation theory.[14]

If memory is made of molecular and cellular substance, and it is transported through the wired systems of its neurological and sensory apparatus, what, then, is the matter of the mental images produced by the mind that we conjure up when reminiscing? The most basic answer coming from a neurobiologist would be that each mindset derived from the brain is made of the same substances: cells, tissues, organs. Due to the mediation of the brain, the mind and its images are grounded in the body proper. The more sophisticated answer, however, includes a refined description of how the mind and consciousness are functions of the brain. Autobiographical memory involves most parts of that well-woven apparatus and comes in various shapes: the recall of facts (where was I born? What is my age?) is as much part of personal remembrance than is the invocation of a familiar mood or event (Do you remember the day your brother was born? How sad I felt when she died!) or the conscious reflection on an earlier stage in life (Have I really changed since the age of 18?). In some instances, memory is an affective feeling that accompanies our seeing a picture or a mental picture we have formed in our minds. To the extent that emotions inform our memories, the stuff of memory may be partly derived from the external object itself (a scenic landscape or a picture thereof) and partly from the construction the brain makes of it (the auditory, visual, tactile, or olfactory perceptions in our minds). Neuroscientist Antonio Damasio calls the latter an "emotionally competent object," referring to the event or object (e.g., seeing a painting, a landscape; hearing a song) that is at the origin of a brain map and elicits a certain feeling: "this picture makes me feel happy" or "this music makes me sad." Invoking a scene or scenery through one's memory may not change the actual object (the painting, the photograph, the record), but it certainly changes the internalized "map" of the initial trigger.[15] Recall and permanent rearrangement of our personal experiences, according to Damasio, play a role in the unfolding of desire. The very desire to re-create an original emotion may be the motivation for changing the brain map: "There is a rich interplay between the object of desire and a wealth of personal memories pertinent to the object—past occasions of desire, past aspirations, and past pleasures, real or imagined."[16]

As neuroscientific research indicates, memory and imagination are not the distant cousins they once seemed: both derive from the same cellular and neurological processes and are intricately intertwined in the matter memories are made of. Memory can be creative in reconstructing the past, just as imagination can be reconstructive in memorizing the present—think only of the many visual tricks people play to perform the cognitive task of factual recall. The function of personal memory, even if restricted to studying its "mindware," is not simply about re-creating an accurate picture of one's past, but it is about creating a mental map of one's past through the lens of the present. The contents of memory are configurations of body states represented in somatosensing maps. Living cells producing this mindware are all but indifferent to the processes they condition, and thus, we could conclude, memory is only the trick the mind plays on the brain. As humans, we even tinker with these processes, for instance by inserting chemical substances (drugs) that alter the body's emotional state. Or, as in the science fiction of *Eternal Sunshine*, technicians artificially remove unpleasant memories by deactivating those neurological circuits responsible for undesirable responses conditioned in brain maps. The erasure of the mental image of an experience in Joel's brain activates a desire to thwart the procedure, which in turn causes the neurological circuits in his brain as well as the circuits in the technician's laptop to go haywire.

Assuming the intrinsic mutability and morphing quality of personal memories laid out by neuroscientists and tested in experimental clinical settings, I now shift the searchlight of this inquiry to a different aspect of memory's matter. In Damasio's as well as most other neuroscientific theories, the nature and materiality of the external object or memory trigger is typically taken for granted.[17] It is obviously not the tangible object they are interested in—the painting, the photograph, the landscape that triggers an emotion or memory—but the contents it represents. Neuroscientists argue the actual pictures become part of the mental maps the brain creates in response to the object, so the materiality of the item does not really matter. But is memory indeed indifferent to the shape and matter of external stimuli and piqued solely by its contents, particularly when it comes to mediated personal memory objects? Is the material artifact that invokes memory irrelevant to mental processes, or does its (changing) materiality have reciprocal effects on the mindware that perceives it? In order to understand personal memory as a complex of physical-mental, material-technological, and sociocultural forces, we may need to understand its distributed matter beyond its embodied nature.

Enabled Memory: "Personal Memory Is in the Mediated Object"

Consider for a moment this all too familiar hypothetical question: What objects would you try to rescue from your house if it were on fire? When confronted with this unwanted yet potential situation, many people rate their shoeboxes filled with pictures, diaries, and similar mediated memory objects over, or on par with, valuable jewelry and identity papers.[18] Whereas the latter two are expendable, the first is considered unique and irreplaceable: memory objects apparently carry an intense material preciousness, although their nominal economic value is negligible. The loss of these items is often equated to the loss of identity, of personal history inscribed in treasured shoebox contents. If you pose the burning house question, asking people whether a mere copy of their original mediated memory objects would suffice, there is a fair chance the answers would be largely negative.

Many of us appreciate these items for more than contents only: we treasure the fading colors on yellowed paper, the fumes of tobacco attached to old diaries, the irritating scratches on self-compiled tapes. Apparently, physical appearances—including smell, looks, taste, and feel—render mediated memory objects somehow precious.

Some cultural theorists have located the matter of memory precisely— and often exclusively—in the tangibility of mediated objects. Walter Benjamin, writing on reproducible memorabilia like personal photographs, called them the "modern relics of nostalgia," the meaning of which lies hidden in the layers of time affecting their appearance.[19] Some contemporary scholars argue that memory materializes primarily through the technology used to produce mediated objects. Media theorist Belinda Barnet, for instance, prefers technology as the main focus for memory research when she writes: "There is no lived memory, no originary, internal experience stored somewhere that corresponds to a certain event in our lives. Memory is entirely reconstructed by the machine of memory, by the process of writing; it retreats into a prosthetic experience, and this experience in turn retreats as we try to locate it. But the important point is this: our perception, and our perception of the past, is merely an experience of the technical substrate."[20] Whereas both Benjamin and Barnet acknowledge that memories actually change over time—one in terms of the object getting older, taking on a sheen of authenticity and invoking nostalgia, the other in terms of technology defining and replacing the very experience of memory—they are adamant in restricting their focus on memory to its material and technical strata only. Barnet argues the primacy of technologies in our production and reproduction of memories. Quite a few mediated memory objects require the original technological apparatus upon later recall because that equipment is indispensable for viewing their contents. Think, for instance, about the projector and roll-down screen needed to show your old slides, an 8-track recorder for playing these antique tapes, or, to stay closer to the present, the hardware and software to read the large floppy disks on which you diligently continued writing your diary after buying your first word processor.

Clearly, the inscription and invocation of personal memory is often contingent on technologies and objects, but unlike Barnet, I locate memory not in the matter of items per se but rather in the items' agency, the way they interact with the mind. Paradoxically, the real value of mediated

objects and their enabling technologies is often thought to lie in their supposedly static meaning, despite their obvious physical decay, and in their
supposed fixity as triggers, despite our constant intervention in their materiality. Memory objects serve as representations of a past or former self,
and their robust materiality seems to guarantee a stable anchor of memory
retrieval—an index to lived experience. But the hypothesis that mediated
memory objects remain constant each time we use them as triggers is
equally fallacious as the outdated theory that memories remain unaffected
upon retrieval—a theory meticulously refuted by neuroscientists. After all,
photo chemicals and ink on paper tend to fade, and home videos lose quality as a result of frequent replay (and even if left unused, their quality deteriorates). In fact, it is exactly this material transformation—its decay or
decomposing—that becomes part of a mutating memory: the growing im-
perfect state of these items connotes continuity between past and present.
Their materiality alters as time passes, but could it be the very combination of material aging and supposed representational inertia that accounts
for their growing emotional value?

Besides a sort of natural physical decay, there is a decisive human
factor in the modification of (external) memory objects. Like human
brains tend to select, reconfigure, and reorder memories upon recall, people also consciously manipulate their memory deposits over time: they destroy pictures, burn their diaries, or simply change the order of pictures in
their photo books. Memory deposits are prone to revision as their owners
continue to dictate their reinterpretation: a grown-up woman ashamed of
her teenage scribbles revises details in her diary; a bitter man erases videos
of his ex-wife; a grandmother takes apart her carefully composed photo album to divide its pictures among her numerous grandchildren.

The double paradox of a stable yet changing external object triggering a stable yet retouched mental image appears all too persistent in our
cultural imagination. In *Eternal Sunshine,* Joel's and Clementine's desire
to destroy their reminders testifies to the human inclination to constantly
revise our past. An endearing scene in Dr. Mierzwiak's waiting room, showing tearful clients holding their bags filled with treasured items to be destroyed, signals the intrinsic modifiability of objects as they are constantly
prone to manipulation and reinterpretation. People have always used material objects not just to store memories but also to alter them, annihilate
them, or reassign meaning to them. Mediated memory objects never stay

put for once and for all: on the contrary, the deposits themselves are *agents* in an ongoing process of memory (re)construction, motivated by desire. Memory allows for both preservation and erasure, and media objects can be manipulated to facilitate and substantiate (new) versions of past experience.

The parallel between neuroscientific theories of memory formation and cultural conjectures of memory is far from coincidental; I dare to argue the two processes intersect. For neuroscientists, the mediation of memory happens in the brain where various interacting neurosensory apparatuses account for their inherent mutability. Cognitive philosophers add to this theory that memories are mediated not only by the intricate brain-mind orchestration but also by the interaction between the brain with physical, external objects it encounters, including the technologies that help make them manifest. Australian philosopher John Sutton, for instance, defines the locus of memory in the hermeneutics of mind and matter; the biggest challenge in analyzing the cognitive life of memory objects is "to acknowledge the diversity of feedback relations between objects and embodied brain."[21] This view is corroborated by Andy Clark, who argues that memory and its enabling technologies are mutually constitutive; he proposes a cognitive science that includes "body and brains" as well as "props and aids (pens, papers, PCs) in which our biological brains learn, mature, and operate."[22] Both Sutton and Clark regard a mutual shaping of the brain/mind and object/technology the inescapable consequence of new neuroscientific insights, and they advocate a concerted interdisciplinary research effort to face the challenge of new paradigms created by these findings.

Indeed, I agree with both philosophers' view that memory is not simply triggered by objects but happens through these objects; brain, mind, technology, and materiality are inextricably intertwined in producing and revising a coherent picture of one's past. However, this double-edged concept can still not fully account for the matter of memory. Memories, in my view, are not only embodied by the brain/mind and enabled by the object/technology, but they are also mediated by the sociocultural practices and forms through which they manifest themselves. Although practices and forms are commonly squarely located in the realm of culture, they cannot be studied separately from the other two conceptual pairs. But before we take that layer further apart into its cultural components, let us first look at how media technologies and objects "matter" as instruments for inscribing personal memory and identity.

Embedded Memory: Personal Memory
as Part of Culture

Mediated memories are material triggers for future recall—produced through media technologies, whether pencil or camera. At the same time and by the same means, however, they are instruments and objects of inscription and communication: devices by which humans seek to establish their own identities in the face of their immediate and larger surroundings. Every historical time frame, as Michel Foucault states, is marked by its idiosyncratic regime of "technologies of truth and self," technologies that "permit individuals to effect by their own means or with the help of others a certain number of operations on their own bodies and souls, thoughts, conduct, and way of being, so as to transform themselves in order to attain a certain state of happiness, purity, wisdom, perfection, or immortality."[23] Instead of attaching technologies of self to the brain, Foucault argues they are always in and of themselves *cultural.* In Stoic culture, for instance, students wrote letters to friends disclosing and examining their conscientious self in order to establish and test their individual independence with regard to the external world. Contemporary variants of former epistolary practices, such as e-mails or weblogs, also help construct a sense of self in connection to an outside world. People (and, one could argue, especially young people) wield media technologies to save lived experiences for future recall and at the same time shape their identities in ritualized processes. We take pictures on vacation for later remembrance but also to convince our friends at home of our relaxed and happy sojourning state; we may want to capture our Thanksgiving dinners on video to document some happy family moments, but a home video concurrently serves to reinforce our notion of belonging to a family. Technologies of self are thus in and of themselves social and cultural tools; they are means of reflection and self-representation as well as of communication.

Foucault's concept may erroneously suggest that media technologies can be regarded apart from their habitual and quotidian use. Naturally, our inclination to take photographs or to write a diary is as much induced by the availability of technologies as by our knowledge of how to use them. As members of a society in a particular historical time frame, individuals deploy a set of practices in common response to their shared social environment and material conditions.[24] Taking pictures, shooting a home movie, or taping recorded music are practices shaped by an internalized cultural logic, unquestioned by one's social surroundings, and performed through seemingly automatic skills. Mediated memory objects provide clues to their social and cultural *function*, thus divulging how people use technologies to produce their own material and representational deposits; these deposits, in turn, betray sociocultural practices.[25] Concretely, a photograph concurrently shows an image and relays information about the habit of taking pictures; a home movie may also reveal something about familial power structures by looking at how various relatives and siblings (often males) take charge of the camera. Some anthropologists even argue that sociocultural practices have their own cognitive properties, which in turn affect (the memory of) individuals.[26]

Besides signifying sociocultural practices, memory objects come in shapes that are often mediated by individual invention in response to cultural convention. Letters or family photographs do not arise out of the blue: we write letters because it is an accepted cultural form. Family albums may literally predispose the kind of photographs we take of our children. Looking at a 1868 photograph of our great-grandfather, we may be touched or puzzled by the stern look of a posing figure eyeing the camera. It is important to acknowledge this memory object to be the result of a historical practice and form: the late-nineteenth-century habit to have a young adult's picture taken by a professional photographer, resulting in a studio portrait. Cultural frameworks are never stable moulds into which we pour our raw experiences to come out as polished products; they are frames through which we structure our thinking and against which we invent new forms of expression.

The significance of sociocultural practice and forms for memory formation poignantly surfaces in *Eternal Sunshine of the Spotless Mind*. For instance, Joel is asked to bring pictures, CDs he bought with Clementine, or tapes they made for each other into Dr. Mierzwiak's office in order to destroy them, because confrontations with these items after his memory's erasure might compromise the procedure's success. Many viewers undoubtedly understand Joel's embarrassment when the technician holds up a mug with Clementine's picture—a commodified form of nostalgia—and empathize with his agony upon seeing him tear up pages filled with sketches and words. When relationships fail, as Joel and Clementine's did, the pain derived from dividing music collections often appears to be inversely proportional to the pleasure of building up communal preferences. More often than not, the collection and consumption of recorded music is a matter of sharing, and the resulting objects are residues of an intensely social process. Compiling tapes with mutually liked music can be an important part of building up a relationship, just as sharing recorded music with others may be a ticket into peer-group culture. Technologies of self are concomitantly technologies of sharing: they help form bonds across private boundaries, tapping into a communal or collective culture that in turn reshapes personal memory and identity.

At this point, adherents of social-economic theory may downplay the relevance of cultural aspects, arguing that memory objects are nothing but products of a commercially induced technology push, promoting new generations of media technologies, forms, and practices for the do-it-yourself memorabilia market (think of the kitsch mug personalized with a picture of your loved one). Following this line of thought would put research squarely into the realm of economics. However, as some sociologists have claimed, the cultural meaning of mediated memory objects and technologies is complex *because* of their inherent linking of private life to public culture. Roger Silverstone, Eric Hirsch, and David Morley, for instance, consider the use of media technologies to be grounded in the very creation of home and "home-ness."[27] A family unit (or household, to use a more economic term) makes a decision about which technologies or media objects enter its private sphere. Studying how these items are appropriated, objectified, and incorporated in the home, the British sociologists try to understand how interrelationships of technology and culture define notions of self and family vis-à-vis society at large.[28] Commercial forces should not be underestimated, but neither should they be singled out as determinant factors in the construction of memory objects.

To summarize my argument so far, scholars from various disciplines have refuted the truism that memories are images of lived experiences stored in the brain that can be recalled without affecting their content. The cliché of (mediated) objects as immutable deposits triggering fixed memories from a mental reservoir is as outdated as the idea of enduring single memories being stored in particular sections of the brain. Scientists and philosophers agree material environments influence the structure and contents of the mind; objects and technology *inform* memory instead of transmitting it. Memory is not exclusively located inside the brain, and hence limited to the interior body, and it cannot be "disembodied," because external bodies and technologies are part of the same mutual affect. To this doubled-edged concept of brain/mind and object/technology I have added a third layer of sociocultural practices/forms that, in my view, complements the other two. Mediated memories perform acts of remembrance and communication at the crossroads of body, matter, and culture.

The Digitization of Personal Memory

Up to this point, I have deliberately focused on cultural memory in the context of analog media technologies. Now that we are entering a digital age, the questions arise: How does digitization change the stuff that memory is made of? Will it modify the "nature" of our brain maps? Does it affect the epistemological or ontological status of digitized media objects? Or will it alter the cultural practices and forms through which we shape our remembrance of things past? Questions like these suggest a deceptive primacy of technology as the impetus for change. History has taught us time and again that a transition from one technological regime to another implies more than the replacement of tools or machineries; it involves a fundamental epistemic overhaul, revising our instruments of living along with our ways of understanding life. Digitization, rather than being a replacement of analog by digital instruments, encompasses everything from redesigning our scientific paradigms probing the mind to readjusting our habitual use of media technologies, and from redefining our notion of memory all the way to substantially revising our concepts of self and society. Obviously, the digital evolution has not changed the "matter" of memory—the mindware enabling conceptions of who we were, are, and want to be—but it certainly affects the way scientists understand the brain performing various functions of memory. And ultimately, I argue, it may change the brain itself, for digitization may impact the brain's constitution just like chemical and genetic evolutions did before.

For one thing, digitization is definitely changing the way neurobiologists envisage and conceptualize memory functions. Activities of the living brain are increasingly visualized with the help of digitized imaging technologies, such as fMRI or positron-emission tomography (PET).[29] An fMRI scanner typically registers specific changes in brain activity: while the person performs a cognitive task, the machine measures the metabolic changes that are linked up with neural activity. In this way, emotions such as fear, aggression, or sexual urge—emotions essential to survival—can be shown to emerge through the paleocortex, the middle brain, whereas the "higher" cognitive and behavioral functions, including reason, are regulated in the neocortex, the outer brain layer.

But what do these scans exactly figure? Do they visualize the matter of memory? Not really. What these machines do is to measure increases in blood flows through the brain; if the brain is more active, it needs more blood and oxygen, resulting in more intensive blood circulation. This activity shows up as red and yellow blots on the screen; representations of the brain at work can subsequently be translated into knowledge about mental processes—that is, by those properly trained in reading them. Medical imaging involves a series of translations in which the body, technology, and expert scientists each play a constitutive role; the beautiful graphics of fMRI, like many techniques heretofore developed in medical diagnostics, imply much more precision, interpretative clarity, and transparency than there actually is.[30] Although the development of fMRI is still in its infancy, the apparatus could theoretically be employed to trace memories as physiological activities in the brain; "trace" is the word here, not "locate," because what neuroscientists actually capture in these images are the changes in neural activity resulting from a specific cognitive, emotive, or conscious task encoded in colored signals. However, most fMRI studies use univariate processing—highlighting only one variable in the brain—a method that shortchanges the distributed nature of neurodynamics. The apparatus tends to confine activity to a specific location in the brain, thus favoring the legitimacy of linking complex mental functions to particular brain regions.

If powerful imaging technologies sway professionals to design research questions supportive of simplified paradigms, it is easy to see how nonexperts are persuaded by the machine's potential to appear like a transparent diagnostic apparatus, linking diseases to exact locations in the brain. Ever since the emergence of X-ray technology, photography has been the dominant model for all kinds of medical imaging—what you see is what you get. It is an increasingly popular tenet—especially in court circles— that complex phenomena like schizophrenia, drug addiction, criminality, or for that matter, "traumatic" personal memories, can show up on digital

scans as pieces of irrefutable evidence.[31] In popular discourse, like magazines and in medical television series, these images often come to stand as
visual proof of a certain diagnosis, be it brain damage or mental abnormality.[32] The movie *Eternal Sunshine* is a case in point: in this Hollywood
fantasy, new digital tools diagnose "ailments" such as traumatic memories
of botched relationships. As said before, the film attests to the latest neuroscientific findings on memory formation; it also plays up to the popular
expectation that personal memories show up in precise spots on the fMRI
scans ("the Clementine files"). Indeed, even if few researchers seriously
believe that brain functions are compartmentalized, the alleged potential
of fMRI machines to visualize "disease" or concrete memories reflects a
strong desire for visual transparency and technological prowess, but is still
far from being a reality.

The diagnostic promise of fMRI, however, is not the real science fiction in *Eternal Sunshine*; the same technologies that help diagnose mental
processes are actually projected to also help doctors intervene in the brain
and thus remedy ailments. Michel Gondry's extrapolation of techno-
experts erasing the Clementine files from the brain in a sort of backward
intervention through the computer may seem a projection on par with
H. G. Wells's time machine. And yet, the image of Joel's head wired and
plugged into the computer—an automatic software pilot deleting memories from his brain—does not look as alien as it should. Why is that?
A couple of explanations may be plausible. First, we have come so used to
technologies depicting our interior body and making it visible on the screen
and translating it into digital code that we begin to understand corporeal
processes as disembodied information.[33] Second, we are rapidly becoming
accustomed to treatments of bodily defects via computers. Computer-
assisted surgery (particularly neurosurgery) is no longer a fictional trope; it
is a fast-developing branch of medicine. Actual bodies are treated from
outside the physiological realm, as surgical interventions mediated through
computers and steered by human hands and brains. And third, already
successful experiments with chemical interventions in the blockage of
traumatic memories in human brains make "informational" interventions
appear far more feasible. However far-fetched it may appear, manipulation
of the mind as a result of computer processing is theoretically feasible. In
fact, brain modification by means of information processing does occur,
albeit more subtly and attenuated than this movie would have us believe.

In many respects, digital imaging technologies turn the brain into a seemingly disembodied, informational entity, and yet it is an illusion to think that memory could be severed from the body, because biology and technology—body and media—have merged beyond distinction. Philosopher of science Eugene Thacker illuminates this process by introducing the concept of "biomedia," defined as the continuous transformation of bodies through the practices of encoding, recoding, and decoding bio-information.[34] Bodies, as Thacker contends, are "both material and immaterial, both biological and informatic."[35] When digitizing the body, the biological is rearticulated as informatic in order to be enhanced or redesigned. Both body and machine are considered platforms through which activities are mediated, yet the materiality of that platform profoundly *matters*: information is embodied as much as flesh is computed. In the long run, the computations carried out by computers will inevitably retool our mindware if only because certain interventions in the body's physiology can not be designed or executed without digital machinery.[36] Of course, it is still a long stretch to prove how this theory applies to memory research in neuroscience, but it is inevitable that digital technologies will impact not only our knowledge of how the brain works but its actual workings.

What we witness in the movie *Eternal Sunshine* is an almost allegorical illustration of Thacker's biomedia, spelled out in a step-by-step encoding, recoding, and decoding sequence. First, Joel's memories of Clementine are translated into digital data—information visualized in fMRI scans. Subsequently, the Clementine files are uploaded into a laptop and "recoded" to be deleted; and finally, Joel's brain is rewired to accept the cleansed "data" into his memory. Joel's resistance of the erasure guys affirms Thacker's contention that informational processes never leave the body untouched and vice versa; the mindware of the brain is not simply retooled by the hardware and software of the computer, but data and flesh are mutually implied in the spiraling process of transformation. The movie delicately— even if awkwardly, in an accelerated compression of time—suggests the inseparability of brain and informatics in its fictional depiction of "digitized memory."

The subtle message behind this movie that computers are both diagnostic imaging tools and instruments of intervention should not be mistaken. Functional MRI scans never take the mind outside the body, just as ultrasounds do not sever a fetus from the maternal womb, and yet they definitely *affect* the body.[37] If we accept the premise that memory is not located either inside the body or outside it in culture but is an embodied experience in which mind, computer, and object form the distributed agency, then the idea of intervention in the brain's function by means of information technology becomes much more realistic. In the long run, not one component in the chain lacing mind, machine, and memory will be left untouched. In fact, brain modification by means of information processing does occur, albeit more slowly and weakly than this movie makes us believe. Technologies of memory are in and of themselves *technologies of affect.*

Digital Memory Objects and Media Technologies

Similar to the myth of disembodiment, digitization often promotes the erroneous presumption of dematerialization. In the first decade of a new millennium, our "technologies of self" are being rapidly replaced by digital instruments, and we are still in the midst of finding out how to adapt to the cultural forms and practices that inevitably come along with this retooling of memory artifacts. What does it mean for personal cultural memory when our tools and objects for producing memories become digital (a term often equated with "immaterial")? What are the consequences of "digitized" objects for our habits of inscribing, storing, and re-creating personal memories? Obviously, digitization carries substantial epistemological and ontological implications, not only with regard to our memory objects and the technologies we use to create them but also with regard to our very *concepts* of memory and experience. Let me briefly elaborate on several of these implications.

In *Eternal Sunshine*, analog mediated memory objects—cassette tapes, framed and laminated pictures, handwritten diary pages—serve as imprints for lost moments; they are the reified items through which we come to know and hold the past, and which need to be destroyed in order to get rid of unwanted memories. The absence of modern digital memory objects (such as digital photographs, weblogs or MP3s) in this movie is rather conspicuous in the face of the fancied digital erasure procedure wielded by the doctor and his technicians.[38] As said before, the supposed fixity of mediated objects has always been illusionary because the very corrosion of analog objects is partial to the "memory sensation." Digital objects, such as photographs, are considered by many to be immaterial because digits are invisible and they can be endlessly manipulated until a final format (printed photograph, music CD) "materializes." However, to understand the digital as immaterial is as erroneous as the idea of analogue mediated objects being static reminders of past experience. Layers of code are definitely material, even if this materiality is different from the analog objects that we are used to and that are still very much part of our personal cultural memory.

Indeed, digital technologies necessitate an adjustment of epistemological horizons: we can no longer assume—if we ever could—a digital photo to accurately represent reality as caught by the camera eye. In many ways, computer memory perfectly suits the morphing nature of human memory over time. Computers are bound to obliterate even the illusion of fixity: a collection of digital data is capable of being reworked to yield endless potentialities of a past. An intermediate layer of coding enables infinite reshaping of pictorial representations of the past before they become manifest in the present.[39] Perhaps not coincidentally, the reconsolidation theory recently adhered by neuroscientists finds its technological and material counterpart in digital media technologies that boost our ability to redesign one's past on the conditions of one's present. The ease of digital manipulation, compared to analog photography, may not just facilitate the airbrushing of images to be stored in our repositories but may also actually augment the role desire has always played in the mental articulation of images, as pointed out by neuroscientists. Personal memories, at the moment of inscription, are prone to wishful thinking, just as memories upon retrieval are vulnerable to reconsolidation. Imagination and memory, in the age of digital technologies, may become even closer relatives.

In addition, the digital condition likely affects the ontological status of memory objects. Memory objects were never immutable items but were always constitutive agents in the act of memory. What changes with the advent of digital cameras, webcams, and blogs in our personal lives is that computerized tools infuse our memory at various stages of the process, and their digital nature (again) probes the boundaries between what constitutes memory and object. The coded layer of digital data is an additional type of materiality, one that is endlessly pliable and can easily be "remediated" into different physical formats. But this new type of materiality is equally vulnerable to decay—a degenerative process that is part and parcel of human memory. The world's computers are brimming over with personal treasures of every genre (music, pictures, texts), but no one guarantees the preservation of electronic materials for generations to come. Machines and software formats may become obsolete, hard drives are anything but robust, and digital files may start to degrade or become indecipherable. Ironically, problems of preservation and access to personal memories, as a result of their digital condition, could become even more complex than before. Even digital memories can fade—their fate determined by their *in silico* conception—as the durability of hard drives, compact disks, and memory sticks has yet to be proven.[40] Memory does matter, perhaps even more so in the digital age.

The Digitization of Culture

Not only does the digital transform brain-imaging techniques and memory objects, but there is also an iterative relationship with the sociocultural practices that inform their use. Whereas in the analog age, photos, cassette tapes, or slides were primarily intended to be shared or stored in the private sphere—a slide show with the neighbors, a forgotten shoebox in Grandpa's attic—the emergence of digital networked tools may reform our habits of presentation and preservation. By nature of their creation, many digital memory items are becoming networked objects, constructed in the commonality of the World Wide Web in constant interaction with other people, even anonymous audiences. Technologies of self are—even more so than before—technologies of sharing. However, the moment of sharing, as a result of the networked condition, may arise much earlier in the memory process; for instance, a photo or diary entry may be sent through the Internet only seconds after it has been made, and it can be distributed among a potentially worldwide audience by a click on the mouse. When it comes to weblogs or MP3 file exchanges, it becomes difficult to describe new sociocultural practices in terms of the old: diary writing or compiling cassette tapes for a friend are succinctly different activities than weblogging or downloading music. Interestingly, people deploy several technologies concurrently when amassing their personal collections; each mediated artifact, whether a cassette tape or MP3 file, not only represents the contents favored at one time in life but also makes a statement about one's preferred mode of recollecting.

Digital cultural forms do not simply replace old forms of analog culture; weblogs only partly overlap with the conventional use of paper diaries, laminated pictures are still printed despite the rise of digital photography, and MP3 files are not exactly replacing our tangible music collections. New practices gradually transform the way we collect, read, look at, or listen to our cherished personal items. The word "gradually" is important to emphasize here, because the ongoing digitization of memory tools and objects all but annihilates analog forms and practices. On the contrary, various theorists emphasize the dynamics of remediation: the way in which new technologies tend to absorb and revamp older forms or genres without completely replacing the old.[41] Photography "remediated" painting, but never took its place, even though both cultural practices repeatedly had to adjust their ontological and epistemological claims in the face of new technologies. Diaries and photo albums are currently undergoing a metamorphosis, although it is hard to predict which status and function familiar paper forms will adopt in conjunction with lifelogs and various kinds of web-based pictorial repositories. In fact, it is likely that analog and digital forms and practices will always coexist, albeit in varying configurations. New hybrid forms and fused practices are likely to inform the larger cultural tendencies that propel their use.

Can we conclude from the above that digitization is, ultimately, a cultural process that is slowly changing the way we remember our selves? The problem with this thesis, as stated earlier, is its deceptive primacy of technology as the cause for change. The matter, nature, and function of memory never changes as a result of technology; rather, the concomitant transition of mind, technology, practices, and forms gradually impinge on our very acts of memory. The first chapter explained how mediated memories manifest themselves along two axes: a horizontal axis expressing relational identities and a vertical axis articulating time. Being active producers and collectors of mediated memories, we carve out our personal niches in the vast sea of culture surrounding us, thus creating a continuum between past and present. In this chapter, I have argued to add a third (diagonal) axis to this model, configuring how memories are mediated through functions of body and mind, technology and materiality, and practices and forms (see Figure 2). Tied in with the horizontal relational axis, it emphasizes how acts and objects of memory are concurrently embodied in individual brains and minds, enabled by instruments and embedded in cultural dynamics. And offset against a vector of time, the model builds in a reflection on transformation.

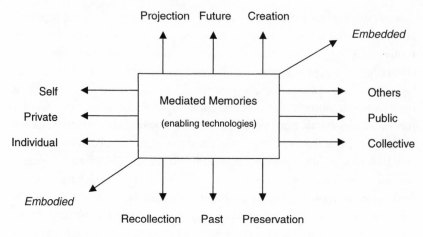

FIGURE 2

Moments of media transition are so interesting because they are periods in which social practices and cultural forms are unsettled and renegotiated—a negotiation that concerns the materiality and embodiment of media technologies as well as the meanings arising from their use. It is at the nexus of mind, technology, and perceptual and semiotic habits that mediated memories are shaped. An insidious process, digitization—conceived as concurrently a technological and sociocultural transformation—is likely to affect our very concepts of memory and remembering.

First of all, the digitization of media may affect physiological and mental functions of memory, as much as mind mechanisms inform our development and use of digital, networked media technologies. Multimedia computers increasingly encompass a divergent variety of personal memory objects and concurrently connect us to a vast network of instantly available visual, auditory, and textual resources. Search engines and digital cut-and-paste techniques allow easy access to, and use of, numerous productions of others—known or unknown, private or public expressions that may or may not invite reciprocity. Memory, as a result, may become less a process of recalling than a topological skill, the ability to locate and identify pieces of culture that identify the place of self in relation to others. The old-fashioned model of the computer as a model for the brain as a means for storage and retrieval may be up for renewal; the computer supports the inherent inclination of memory to store *and* revise, to download *and* upload, to recollect *and* project or invent. Of course, memory was always a creative act that involved communication as much as reflection, and yet it remains challenging to analyze and identify concrete instances of how mental processes are implicated in a larger pattern of transformation.

Another profound change in the transformation from analog to digital lies in the emergence of multimedial, multimodal technologies, objects, forms, and practices. If we look at our analog mediated memory objects, they commonly fit single categories of media and perceptual modes. For instance, a diary used to be a paper object that favored writing (despite the occasional drawing or illustration); a photo album contains laminated pictures (although occasionally annotated by handwritten comments); and a compiled cassette tape caters to our auditory dimensions of memory. In the digital era, it becomes easier to tie in a single memory object with multiple modes and media. The weblog is no longer strictly a piece of (hand)writing, as the incorporation of music and picture files expands the possibilities of computer-mediated reflection. Digital cameras carry standard options of adding verbal tags and allow the shooting of moving images, and the MP3 player appears to smooth the revival of audiobooks. The multimedial and multimodal potential of digitization is not merely an interesting side effect of technology but may ultimately redefine the sensory ways in which we catch and store memories. Visual, auditory, and verbal memory objects are not confined by the sensory mode inscribed in their enabling media; instead, mediated memories may become an intrinsically multimodal reservoir for creative inventions. Hence, diary writing may no longer be "a matter of script"—an utterance contained by its material and technological parameters—but could yield innovative ways of expressing the multimodal self.

Finally, science imaging and technological imagineering are inseparable from the forces of cultural imagination. While we are reinventing the tools for remembering, fantasies of digitized memories enter our popular culture. Technologies of self are intimately interwoven with cultural products: home movies, for instance, surface in Hollywood blockbusters, family albums become online multimedia productions, and tape collections inspire grand-scale schemes of music swapping. Future memory objects and acts of memory may be produced digitally, but they will be inevitably shaped by desires and concepts previously developed in the era of chemical, magnetic, and mechanical reproduction. *Eternal Sunshine of the Spotless Mind*, released at a moment in history when personal memory finds itself caught between analog and digital materiality, helps us reinvent the meaning and function of personal memory: What do we expect and want from our new tools? How do we envision the role of memory in our lives and how would we like to change it? The invention of every new technology—whether photography, video, or the Internet—revises our methods of personal remembrance, and each of these same tools influences the way we imagine and inscribe our selves in relation to the culture at large. In fact, this may answer the question *why* memories matter: humans have a vested interest in surviving, and therefore they invest in creating and preserving imprints of themselves—their thoughts, appearances, voices, feelings, and ideas. They may want these images to be truthful or ideal, realistic or endearing, but most of all, they want to *be* remembered.

The proposed model is intended as an analytical tool; it serves as a model for understanding complexity, reminding us of the intricate multifaceted, interdisciplinary, and dynamic nature of memory. Mediated memories, in their conceptual and material dimensions, are always in transition; they are infused with technology and yet always also embodied and enculturated. The remainder of this book puts the analytical power of this model to a test. In each of the next four chapters, concrete mediated memories (diaries, sound recordings, photographs, and videos) form the lens through which to examine a specific aggregate of minds, objects, technologies, forms, and practices at this transitional stage between analog and digital. Acknowledging the changing epistemological and ontological status of mediated memories, I explore how digital culture is revamping our very concepts of memory and experience, of individuality and collectivity, unsettling the boundaries between private and public culture in the process. Digital technologies, which are part of a culture whose cognitive and epistemic paradigms are under construction, are as much reflections as they are agents of change. Personal cultural memory is coming out of the shoebox and becoming part of a global digital culture—a wireless world that appears dense with invisible threads connecting mind, matter, and imagination.

Writing the Self

When I was a teenager, I kept a diary in which I occasionally used to express outbursts of intense anger or passion—the kind of emotions average teens typically want to get off their chests. One day, I remember writing a disgruntled poem about my parents who allegedly policed my freedom by setting unreasonable curfews; I wrote several drafts before trusting the final words to my personal notebook, which I then, quasi-casually, left open on my desk for my mother to read. The next day after school, I found a subtle note stating that gutsy teens may one day grow up to become worried parents. Obviously, my words had resonated with their intended reader, who had responded in kind. But when my older brother, later that evening, teasingly joked about a new poet in our family, I was totally embarrassed and indignantly accused him of privacy violation. His defense that thoughts meant to be kept to oneself should be locked up and words intended to be read by one specific person should be properly addressed appeared difficult to rebut. After all, I had been the one who had intentionally left my scribbles open for another pair of eyes, even if I never meant those eyes to be my brother's.

This memory occurred to me when recently stumbling upon a newspaper story about two female students who found themselves exposed after their written evaluations of some boys' sexual prowess had been distributed to their classmates (and beyond that circle). The two young women had communicated their intimate thoughts via a so-called lifelog, but they had apparently misjudged or neglected the fact that their diary entries were also accessible to other readers. It did not take long for their peers to trace their revelations and the male students' revenge was more than a little awkward for the authors, who presumed their readership to be limited to one friend only. Like my own diary, the teenagers' lifelog was directed to one other pair of eyes but it was unwittingly exposed to the public eye.

The paper diary and the electronic weblog have much in common in terms of function and use. For many centuries, the diary has been characterized as a (hand)written document that chronicles the experiences, observations, and reflections of a single person at the moment of inscription. The diary as a cultural form is varied and heterogeneous, yet it is typically thought to represent the record of a single person constructing an autobiographical account. Inviting the translation of thoughts into words via the technologies of pen and paper, the diary symbolizes a haven for a person's most private thoughts, even if published in print later on. Diary writing, as a quotidian cultural practice, involves both reflection and expression, but it is also a peculiarly hybrid act of remembrance and communication, always intended for private use, yet often betraying an awareness of its potential to be read by others.

A similar ambiguity can be found in what is considered the diary's digital equivalent. Since 1995, weblogs have become a popular genre on the Internet, as millions of people are now heavily engaged in blogging— expressing and exchanging their personal accounts through the technologies of hardware and software.[1] The term "weblog" or "blog" is a rather general label for a variety of interactive forms; of all weblog variations, the so-called lifelog seems to come closest to the traditional diary genre.[2] Lifelogs are multimedial online experiments in self-expression, but they are often cross-linked to other lifelogs, creating blog communities.[3] Searching on the Internet today, one can find a plethora of digital diaries, everything from travel blogs chronicling the climbing of Mount Everest to personal blogs documenting the spiritual journey of a born-again Christian, and from the intimate exchange of sexual experiences between teenagers to the shared stories of earthquake victims.

How do old-fashioned paper diaries compare with digital lifelogs? My own experience and the experience of the two female students appear strikingly similar in showing how diary writing and blogging are both ambiguous acts, rife with ambiguities concerning privacy and publicness. And yet, the lifelog is not simply a contemporary replacement of the diary. As mediated memories, diaries and lifelogs move along the axes of relational identity and time: they are instruments of self-formation as well as vehicles of connection. They are also tools to record and update the past that simultaneously steer future memory and identity. In order to understand how diary writing and blogging have evolved, we need to look at them as acts involving mind and body that are enabled through technology and inscribed in social practices and cultural forms.

In the next section, I look at how diaries and lifelogs are deployed to scaffold individual memory, funnel emotion, and create affect. The Internet-based tool has recently been mobilized in Alzheimer's disease patients' struggles against forgetting—to preserve a sense of self in the face of a harrowing degenerative disease and connect to fellow patients. The third section shows how the digital materiality of the Internet engenders a new type of discursive awareness that is both a continuation and a substantial change from the days of pen and paper.[4] In their lifelogs, teenagers play with material limitations and technological potentials: privacy and intimacy are not so much social conditions as they are rhetorical and communicative effects. Therefore, it is critical to also consider diary writing and blogging as sociocultural practices and explore how preferred ways of expressing the evolving self change in conjunction with cultural conditions. Tracing diary writing at a time of cultural and technological transition prompts us to reflect on the epistemologies of private versus public and of recollection versus projection.

Diaries, Lifelogs, and Affective Subjectivity

Diary writing is often considered the discursive resonance of a person's subjective emotions, channeled by the mere jackets of words and grammar and trusted to a piece of paper or a screen that has no other function than to mirror its contents to the writer. However, this interpretation is too simple as it ignores the mental complexity involved in diary writing. A person's subjectivity, neuroscientists contend, develops through articulating one's intimate feelings in image maps and through language; the mind constitutes a notion of self, and subjective experience is shaped by various means of expression—a process in which memory and imagination play equally large roles.[5] The mind and its consciousness are of course private phenomena, and as much as they offer public signs of their existence, these utterances should not be equated to subjective experiences processed by the mind. When individuals express their intimate thoughts, these verbal articulations are interlaced with behavioral, social, and cultural aspects of what we call enunciation. Written records of personal experience are always mitigated by the tools and conventions of writing. In order to deconstruct the myth of diaries forming a verbal template for the outpour of reflections and feelings, we need to look more closely at the mental process involving subjectivity and affectivity.

American psychologist Silvan Tomkins has argued that the emotional formation of the subject occurs not in isolation but happens in complex interactions or "affects" with others and with thought; affects, according to Tomkins, are "neuro-physiological events" that are part of a larger cognitive system operating as a series of distributed functions, including sensory perception and memory.[6] The affective constitution of personal memories is well recognized by psychologists: when people read or hear reminiscences narrated by others, they often feel triggered or invited to contribute their own memories. Based on Tomkins' theory, Australian psychologist Anna Gibbs labels this phenomenon "contagious feelings," which she describes as a process of "neural firing," eliciting a positive feedback loop "in which more of the same affect will be evoked in both the person experiencing the affect and in the observer (a phenomenon known as 'affective resonance')."[7] In other words, subjectivity and affectivity constitute each other in a constant feedback loop between self and others, where the narration of experiences, memories, and feelings of others contribute to the formation of self. Media, as Gibbs contends, introduce a powerful new element in this state of affairs: they literally and metaphorically act as amplifiers of affect while dramatically increasing rapidity of communication and audience reach. Mediated affect may thus at any point add to the feedback loop of affective subjectivity: personal stories told on camera and broadcast by television may trigger associated feelings and memories in individual viewers, contributing to that subject's emotional formation.

Gibbs implicitly transfers the notion of affective subjectivity from autobiographical memory onto cultural memory by appointing contemporary electronic media as the amplifiers of affect: stories told on television, radio, or the Internet encourage public exchange of intimate feelings and emotions. But the mediated affect she describes did not originate with the advent of electronic media such as television; for centuries, paper diaries have provided this function. Diaries or personal notebooks, as the material signifier of subjective reflection, never have been simply expressions of the individual's emotional state or psyche but are themselves cultural forms that trigger affective response. Needless to say, there is no single, uniform definition of what a diary is, and over the past centuries, its classification has been anything but homogeneous.[8] The diary obviously finds its roots in the daily recording of events, transactions, feelings, and reflections; in contrast to the journal, the diary is commonly referred to as a private kind of reflective writing produced by a single author and closed to public scrutiny. In general, the taxonomy of the old-fashioned paper diary tends to be based either on its contents (personal, intimate self-expressions as opposed to daily records of fact) or on its directionality (intended for private reading versus public use).[9] However, this opposition is fallacious.

If we closely look at how paper diaries have been used in the past, the characterization of diaries as enunciations written strictly for oneself is hardly tenable. Keeping a diary is at once a creative and a communicative act, and it also serves as a memory tool: writing the self constructs continuity between past and present while keeping an eye on the future. Reading one's own scribbles at a later moment understandably elicits a tendency to rework hindsight experiences into one's personal reflection and to edit the original entries not only for grammar and punctuation but also for content. The diary's contents, when reread at a later stage in life, may either yield nostalgic yearning or retroactive embarrassment, in some cases even leading to a definitive destruction of the object. Anne Frank's famous diary is a case in point: she began revising the first version of her personal account in March of 1944, several months before she was deported to a concentration camp, inspired by a government's official radio address, urging Dutch citizens suffering under Nazi occupation to keep personal notebooks for later publication. The various drafts have now become an essential part of her legacy: they show how keen she was to revise her diary to include hindsight observations just a year after writing her first draft, as evidence of her evolving or growing self. Equally important in this regard is the fact that Anne never intended to keep the diary strictly to her (future) self: she wanted to become a published author. Anne Frank's desire to share experiences and evoke affects with (potential) others was very much at the heart of her effort. Affective subjectivity is thus embodied in the *Diary of Anne Frank*, the published version whose mediated affect has inspired many teenagers and young adults to articulate their own personal struggles of coming of age in the face of oppression or diaspora.

A similar amalgamation of subjectivity and affectivity can be discerned in online blogging, although in a distinctly different form. Specifying the mediated affect of contemporary blogs, I turn to a specific type of lifelogs produced by patients suffering from Alzheimer's disease (AD).[10] By means of therapy, patients suffering from dementia and AD have long been encouraged to try to retain a sense of self by looking at old photographs and rereading old notebooks and letters. Supported by relatives and friends who want them to remember their (shared) pasts, one of the most frustrating problems these patients face is their growing inability to recognize themselves and their loved ones as depicted in pictures or described in words. Adding to this frustration is the increasing distance, due to memory loss, between a patient's mental world and the (social) world surrounding them. To an AD patient, the past is disconnected from the present, and often relatives inevitably lose track of the patient's mindset, thus severing ties between the worlds of self and others. Some patients, particularly those coping with an early onset of the disease, have begun to explore lifelogs as a new means to express themselves and share their experience with others. Upon his own confrontation with the disease, Morris Friedell, a retired professor from California, cofounded the Dementia and Alzheimer's Support Network International (DASNI) and initiated a blog community of AD patients in 1999.[11]

Looking at various AD patients' lifelogs, their most observable role is in the sustenance of subjectivity. Some patients poignantly describe their first symbolic confrontation with their "new self": they find themselves staring at the evidence of cerebral atrophy in the diagnostic scans made in the hospital, while the doctor explains the signified meanings of the slices on the screen. The look into the brain often leads to confusion and discussion. As Friedell discovers, brain scans (PET, fMRI) are often indecisive and ambiguous; a discussion on the Internet forum may help patients understand what medical images mean in terms of diagnosis and prospects.[12] Once the initial shock of their confrontation with a shrinking brain volume subsides, patients discover the blog as a tool for self-preservation; as the disease progresses, it becomes vital to document what can still be remembered and what is lost to oblivion, hence keeping track of a changing sense of self. To some extent, the lifelog functions as a personal memory aid in which quotidian chores and events are recorded before they are literally erased from the mind. But in other ways, the lifelog becomes a technology of affect: through links included in each lifelog, patients become part of a blog community of fellow AD patients who may just read about each others' ordeals or share experiences.[13] Most entries in AD patients' blogs are written with different readers in mind—fellow patients, family members, or everyone who is interested in their private struggle with the disease.

Chip Gerber, an AD patient from Ohio who was confronted with symptoms a few years before retiring as a social worker, keeps a blog of his everyday experiences and hopes. He invites everyone to join him on his involuntary journey as he enters "the long good-bye." In one of his entries, Gerber relates how he forgot how a toaster works and subsequently burned himself. He expounds: "Those of us with dementia do some funny things at times. For me it is a time to laugh and make a joke out of it. It's going to happen again in one form or another. It makes a good story to share with others . . . if I remember what happened, that is. Then others can enjoy the event with me also. A sense of humor is a gift that I've been given and I highly recommend it to others."[14] The awkwardness of such moments of forgetting comes across as particularly striking, not only because they are apparently shared with friends and family—the intended effect of intimacy—but also because the author directs his entries to other patients, encouraging them to remain positive in the face of their unnerving mental deterioration.

As they keep their blogs and link entries to those of fellow patients, a community emerges that embraces the blog as its connecting medium. This affect is powerfully amplified when outside readers like me realize the blog serves as a personal marker of cognitive abilities the patient will inevitably lose at a later stage. It is thus the prolific, yet ambiguous combination of writing to oneself and to others that constitutes the affective subjectivity in Gerber's lifelog.

Most of the weblogs of AD patients are interlinked and connected, and so it is clear that the patients regularly take notice of other's contributions to the blog community. Affective subjectivity appears to be contagious, to use Anna Gibbs' term, as one articulation of emotion triggers the next one, which in turn adds to the communal effort. From this perspective, blogging becomes a sort of feedback loop in which subjectivity and affect work reciprocally to constitute the formation of self in constant interaction with others. And yet, the need to express oneself as an individual and train the brain's cognitive functions never appears to be at odds with the need to connect and communicate. Consider, for instance, the blog of Mary Lockhardt, an Oklahoma woman who ran a licensed day-care home for infants until she was diagnosed with Alzheimer's at age 55. She had to give up working and started her blog in addition to opening a chat room for patients like her. These online activities, according to Lockhart, helped redirect her purpose in life: "This is a very lonely disease. I don't want anyone to feel as lonely as I did when I was diagnosed (I thought I was the only one over 50 that had Alzheimer's). If I can help one person with my blog it will all be worth it."[15] Bound together by the same ordeal—people "who are in the same leaky boat," as AD-patient Morris Friedell phrases it—each blogger trusts personal emotions to this novel medium and shares his or her experience with relative strangers who are held captive by the same neurological conditions. Yet the fact that this is a shared weblog does not keep Mary Lockhart, or other bloggers, from listing the birthdays of relatives and friends and reporting how she celebrated each of them. These comments read like a typical boring diary entry—until you realize, as an outside reader, that remembering birthdays is a great struggle in the life of a person coping with dementia or Alzheimer's.

The hybrid notion of the individual yet collective nature of remembrance is crucial to the understanding of why and how lifelogs work the way they do. Australian media theorist Robert Payne argues how the online dynamics of collective recollection stimulates the articulation and rearticulation of the self. Analyzing an Internet-based collective project named Random Access Memory, Payne relates how reading through other people's assorted memories—organized by themes or years—activated him to trust his own personal memories to the screen, thus contributing to the overall project and in turn stimulating others to revise or reenact their memories in narrated form.[16] Random Access Memory, according to Payne, forms the online materialization of "interaffectivity": an exchange of individual stories making a community of records.[17] Of course, Random Access Memory entries do not have the same function as the entries posted by Alzheimer's patients in their lifelogs, but they equally enhance people's inclination to (re)construct the self in the light of experiences posted by others.

For Alzheimer's patients, blogs appear to be the ultimate tool to comprehend and preserve their unique sense of self, if only because they can go back to what they wrote previously; patients constantly fine-tune their own mental and cognitive perceptions of self in the wake of ordeals described by others. The most debilitating aspect of AD and dementia is that patients lose their sense of time, blurring past and future. Each day in the present can confront them with a different persona that strikes them as self, but a self that has lost a substantial part of his or her long-term and even working memory. The most touching parts of these blogs have to do with patient's idea of continuity—how past and future connect in the present understanding of their fate. In Gerber's weblog, for instance, he describes how he and his wife have prepared for the final good-bye: "It is so easy to put off unpleasant tasks, such as doing one's will, getting a durable power of attorney, making funeral arrangements, living wills, visiting an elder law attorney, making one's final wishes known and putting them in writing. The death of a person is not something individuals like to dwell on."[18] Or, as Friedell phrases it, the ultimate goal of his digital effort to record his changing personality is less to guarantee his own continuity as a person than to be a sort of discursive continuity in his family: "Ultimately, I want to leave behind a message for my descendents that when life pitches them a curve they don't have to curl up and fade away. They can get back on their feet and continue being the persons the Source of Life meant them to be."[19] Besides countering the loss of memory, the purpose of these and many other lifelogs kept by Alzheimer's patients is to retain a sense of continuity in present and future—a notion crucial to autobiographical memory. If they relied solely on old photographs, letters, or stories told by others about their pasts, AD patients would hang on to a static notion of self. By keeping a lifelog, they recognize how the disease influences subjective judgments, bearing witness to this change by writing about it in their logs. The strategies of affective subjectivity and interaffectivity help them to share a very individual mental experience and at the same time let other people in on their lonely journeys. Mediated strategies of effect and affect are deployed to keep continuity between their past and current environment.

The similarities between paper diaries and electronic lifelogs, when it comes to the double focus of expressing individual emotion and evoking affect, are striking; their historical continuity is also palpable in their functions to mediate between past and future. And yet, networked lifelogs are distinctly different in that their default mode seems to be connectivity and communication, as opposed to the default mode of diaries, which is isolation and reflection. Notions of subjectivity and connectivity are equally important in AD patients' fight against memory fading and isolation. In addition, the Internet-based tool appears to diffuse the boundaries between privacy and publicness. Whereas Anne Frank's intimate reflections took a long time to move from private diary to public memory object, the built-in potential of lifelogs to open up one's writings to selected audiences changes their intent and impact. To further explore these changes, we need to turn to the technological and material aspects of diaries and lifelogs.

The Technology and Materiality of Diary Writing versus Blogging

Diaries are commonly valued for their contents rather than for their look or feel. But the materiality of diaries and the technologies by which they have come into being are crucial factors in their signification. When referring to paper diaries, two typical concepts spring to mind: the empty diary, preformatted for daily use, which we can buy at stationary stores; and the original, handwritten manuscripts of diaries that have later appeared in print and became widely read. The physical appearance of a prefab diary prefigures the functions of its intended use: empty pages, with or without lines, bound or unbound, dated or undated, offer the author stimuli to fill the more or less blank surface with personal inscriptions and thoughts. In some cases, the diary is completed by a lock and key—a potent symbol of its private nature. The preformatted diary has always been, to some extent, a product of contemporary fashion, its design and layout representing a particular style and catering to a specific age or taste. A diary's materiality forms an essential part of its content. Over the years, diarists often grow fond of the look, feel, and smell of their notebooks; fading paper, youthful handwriting, and ink blobs trigger reminiscences the way photographs do.

Arguably, diary writing is not necessarily inspired by prefab formats: many diaries that were discovered many years after they were written and subsequently published in print had first been scribbled onto single sheets or in simple notebooks. The actual manuscript of such a diary, its original inscription, becomes a vital sign of authenticity—often stored in special places and only accessible to owners or researchers. In the case of Anne Frank's diary, written partly in notebooks and partly on single sheets of paper, the written pages have become the centerpiece of the Dutch teenager's legacy. The manuscript, stored in Amsterdam, appeared in such demand that the Anne Frank Foundation had two exact duplicates made: one to replace the original on display at the museum, the other to satisfy the many requests from film media directors, researchers, and documentary makers for pictures of the original. The original manuscript was locked in a safe place to protect it from further deterioration. Its faithful copies concurrently underscored the idea of the diary as a public commodity and withdrew the original's unique materiality from the media's or public's eye. The manuscript's materiality constitutes an intricate part of the diary's genesis and forms the stake in controversial claims to its authenticity, (uncensored) originality, and completeness.[20]

Pivotal to the materiality of diaries, up to the age of computers, has been the notion of script: the concept of the diary is commonly associated with (hand)writing, signifying not just authenticity but also personality. Handwriting has historically been believed to betray the corporeality of its producer—graphology being the study that yields clues to the writer's character, such as age and even personality traits. Regarded as the first technologizing of the word, the availability of pen and paper facilitated the need to make oneself legible to the other or to the future self. Writing is often tied in with a stage in one's personal development: a teenager's scrawls betray his or her inexperience with the prime tool of literacy—the immaturity of body or mind.[21] Sonja Neef, a German media scholar, claims handwriting is an embodied practice enabled by technology: moving a pen onto paper involves a direct connection between body and script, an act in which the eye and hand are intimately interwoven with the technology of paper and pen and the techniques of deploying them; the hand—a body part instrumental to the "Verkörperung" (embodiment) of thoughts—fixes the inner self to the outside world.[22]

As our technologies for writing change, so do our ways of creating self-reflective records; memory, in other words, is always implicated in the act and technology of writing. Handwritten diaries as material artifacts are themselves memorials—traces of a past self. When Sigmund Freud wrote his essay "A Note Upon the 'Mystic Writing Pad,'" in 1925, he regarded writing and technology as external aids or supplements to memory. Freud described memory in terms of writing, comparing it to the surface of a writing pad that allowed the scribbling of endless notes that could subsequently be erased and yet remain stored in the subconscious layers of the pad, below its material surface. Jacques Derrida, commenting upon Freud's essay, dismisses his notion of writing as an external memory and emphasizes instead technology's instrumental relationship to language and representation.[23] Technologies, including writing utensils, are machines that engender representations while infiltrating agency; pen and paper, therefore, produce different modes of writing than the typewriter or the word processor. Handwriting never simply structures reflections or thoughts but literally creates them; by the same token, a typewriter constitutes a different relation between author, words, and representation. It may not be a coincidence that typewriters never became popular in connection to diary writing; unlike handwriting, the noise of fingers pounding on a machine severed the physical intimacy between body and word.[24]

The advent of the stand-alone word processor, as the successor of the (electronic) typewriter, further disembodied the production of written language, because not only the keyboard but also the screen interfered with the continuity between hand and words. Yet two essential features of word processing may have restored some of the intimacy lost with the typewriter. First, the relative silence of word processors refurbished part of the quietude inherent to solitary writing, and the technology sped up the production of text and maintained standardized letter output. Even more profound has been the ability of word processors to produce tentative texts, provisional versions of thoughts, forever amenable to changes of mind; the editing of visualized words does not leave a trace in the ultimate print. Words on the screen, stored in digital memory, thus formed a new stage in the trajectory between immaterial thoughts and textual products, allowing for invisible revisionist interferences in one's memory. On top of that, digital files may never materialize into print, and they can remain stored in the black box of a personal computer without ever being erased or retrieved (by the writer or by others). Diaries produced by a word processor, therefore, are fundamentally different from diaries produced by means of handwriting or typewriters: the personal computer provides a textual paintbrush that allows editing of one's personal records without leaving a trace. The potential of digital editing at a later stage dilutes the concept of the diary as a material, authentic artifact, inscribed in time and on paper.[25]

In the late 1990s, when stand-alone word processors gradually gave way to networked computers and the Internet became a popular medium for interaction, the physical artifact of the diary seemed incommensurable with the prime demands of instant, ubiquitous connection rooted in digital materiality; mutatis mutandis, the evanescence of the Internet appears at odds with the diary genre's preference for a fixed material output. Moreover, the time lapse between the writing and potential publishing of the diary in print contrasts the immediacy and connectedness of the electronic superhighway. Between the body and the printed word there is no longer just a piece of paper but an intricate technological network of connected individuals and communities; between private thoughts and published words there are only a few seconds before the inscriptions enter the virtual realm. And yet, perhaps surprisingly, we still consider the lifelog to be a digital descendant of the paper diary, except that there is no printed output, only a screen-based one. How, then, should we consider the new materiality of the lifelog? Since computers do not emit odors, and the screen has no particular feel, how can we define what the digital matter of lifelogs actually consists of?

Analogous to the preformatted paper diary and the diarist's handwriting, we can locate the materiality of lifelogs—apart from the obvious hardware of the computer—in two different areas: weblog software and the signature of its users. The first weblogs were operated mostly by digirati, but as specially developed software made blogging technically easy, more people without any specific technological skills joined the various kinds of blogging groups. Since the year 2000, a large number of software packages have flooded the market, enabling even the clumsiest person to become a sophisticated blogger. Software-mediated blogs are overwhelmingly popular among teenagers between thirteen and seventeen years of age who are interested in both lifelogs and scrapbooks. They can choose from a variety of different packages; besides open diaries on the web, such as Opendiary .com and MyDearDiary.com, there are also weblog services for which you need to sign up or even be introduced by a member, like LiveJournal, Blurty, Xanga, DeadJournal, Blogger, and DiaryLand. Although they all basically serve the same purpose, the formats may differ in layout and digital possibilities. To some extent, these different designs resemble the preformatted paper diaries for sale at stationary stores. It seems like the various software formats attract different audiences, catering to heterogeneous tastes and lifestyles, much like brand names of fashion products. As Emily Nussbaum points out in her ethnographic, journalistic report of bloggers, the formats vary only slightly, but users' choice of software packages may be based on particular technical features.[26] For instance, websites like Xanga offer the possibility to give "aProps," a kind of gold star rating for particularly good posts, whereas LiveJournal allows for selecting features such as "current mood music" and "embedded polls or surveys." In a way, blog software resembles the preference for jeans: nuances in style and brand name are important to individuals who seek to belong to a group.

Digital weblogs, in terms of their materiality, may not even remotely resemble their paper precursor, but there is still a distinct continuity in their personal signature. If handwriting betrays a diary writer's character and level of maturity, the typewriter and later the word processor had already erased that trademark of personality; and yet, through word choice, style, punctuation, and the use of emoticons it is remarkable how much lifelog entries give away a person's character. On top of that, the personality of a diarist is even more traceable through his or her prolific choices of cultural contents. A blogger may attach references to (or actually attach) songs, pictures, movies, books, and so forth. Despite prescriptive software formats, weblogs offer a relatively high degree of creative freedom; users can cut and paste from all kinds of media sources, thus exploring their taste by testing it against the taste of others. Weblog architecture encourages this type of self-exploration through encoded features such as "my current mood," "mini biography," and "my interests."

The most profound difference between a paper diary and a blog may be found in the blog's networked condition; lifelogs, by default, are technologies of affect—connecting the individual to others, whether selected groups or an anonymous audience. Some websites offering lifelog software (like OpenDiary.com) allow users to search entries by age group, gender, theme of the week, subject, and cultural preferences; standard features such as "friend groups," "syndication," and "communities" (LiveJournal), prefigure a connected exploration of a user's personal life. What the Internet does best is to create a forum for collective discourses. In this culture of sharing, the lifelog finds its natural habitat: the digital diary becomes instrumental as its multimedial modality equally allows for the creation of one's personal entries as well as for the exchange of cultural contents (clippings, files, songs). Blogging software and Internet hardware, in this argument, are neither neutral technical conduits nor simple commodities, but they are technological tools facilitating a social process in which exchange and participation are conditions of enacting citizenship.[27] The flipside of the culture of reciprocity is instant marketability: personal taste and cultural choices become instantaneously traceable and marketable to commercial ventures. Whereas many diaries (like OpenDiary.com and DearDiary.com) started out as small communities of like-minded individuals, many of these services have been bought by corporations.[28] In a networked environment, where information is constantly cached, weblogs have become gold mines for data diggers. Nostalgic notions of personality still persist, even if new media prompt a keen awareness of technological strategies directing individual taste and community building.

Although the multimedial lifelog appears very different from the preformatted lock-and-key paper diary, each materiality gives away distinctive clues to an author's subjectivity and personality. Just as paper diaries reflect someone's age, taste, and preference at a particular moment in one's life, the software and signature of blogs seem to accommodate the needs of especially contemporary teens and young adults to express and sort out their identity in an increasingly wired, mediated world. Digital technologies are imperative to the creation of blogs, but technology does not tell the whole story; in our focus on technologies, we often tend to understate the importance of their uses. In conjunction to the technological script, we hence need to look at how social practices and cultural forms transpire through the concrete manifestations of diary writing and lifelogs.

Diaries as Cultural Forms and Blogging as a Social Practice

The myth that the diary is a private object, written strictly for oneself, contrasting a journal of fact, which is written for others, is as misleading as it is persistent. As literary scholar Thomas Mallon argues, no one ever kept a diary just for him- or herself; pointing out the continuity between the journal and the diary, he concludes that both are directed toward an audience and "both [are] rooted in the idea of dailiness, but perhaps because of the journal's link to the newspaper trade and the diary's etymological kinship to 'dear,' the latter seems more intimate than the former."[29] Of all the varieties within the diary genre, some are written more with a reader in mind than others, but an essential feature of all diaries is their (imagined) addressee. Some authors direct their diaries to God, to an imagined friend—like Anne Frank's "Kitty" or André Gide's mysterious addressee—or to the world at large; in any case, the notion of addressing is crucial to the recognition of diary writing as an act of communication.[30] Even as a form of self-expression, diary writing signals the need to connect, either to someone or something else or to oneself later in life.[31]

Another misperception we can trace in the genealogy of diaries concerns their authorship. They were commonly ascribed individual voice and authorship—supposedly written not only for one person but also by one person. However, there are a number of historical examples in which the genre was deployed as a communal means of expressing and remembering. Many religious congregations, for instance, considered the diary a semi-public record, shared within but never outside a community.[32] In the history of diary writing, the genre as a communal means of expression has found many practitioners, from South Pole explorers to POWs held in captivity.[33] For groups bound together by an adventurous ordeal, a collaborative diary was often a means to trust one's personal emotions to a relatively safe medium and share the experience with mates held captive under the same conditions.

We often encounter preconceived notions of the diary as a private object written by a single author when it is contrasted to the lifelog; the digital networked variant is thought to privilege connection over individual expression and group formation over private authorship. Indeed, exploring one's identity in the various scenarios allowed by online groups is part and parcel of experimenting with multiple aspects of a personality, specifically among teenaged bloggers.[34] Many young bloggers engage in diary writing to be seen or to see themselves through the lenses of others—a kind of validating experience even if no one else ever reads the diary. And as we witnessed in the blogs fostered by AD patients, they obviously value connectedness: the potential to address others. Connectivity and sharing apparently form the default mode of blogging, as opposed to individual expression and monologue, which seem to form the default mode of paper diaries. And yet, as argued above, neither the historical genre of the diary nor its contemporary digital variant fits this binary genre frame: the functions of self-expression and communication have never been at odds but have always been (and still are) copresent in the genre. Diaries have historically been produced as a sort of materialized feedback loop—dialogic rather than monologist—in which subjective emotions trigger responses that contribute to an individual's formation. Lifelogs continue to espouse this hybridity, which is less a feature of the genre itself as it is an effect of its cultural use.

From our reading of historical diaries, we may deduce many details concerning the social practice of diary keeping; by the same token, we can tell from reading online lifelogs (complemented by ethnographic research) how contemporary teenagers deploy the new technologies to perform daily rites. Both diary writing and blogging are ritualized activities that gradually receive a place in a person's life. Writing a diary, of course, never happened in a social vacuum; the ritual occupies its own niche alongside other acts of communication, such as talking, listening, reading, and so forth. As a quotidian habit, diary keeping gives meaning and structure to someone's life, and therefore it is important to regard this activity in the context of everyday activities. In the case of Anne Frank, writing a journal created a zone of silence and refuge in a small space, densely crowded and heavily trafficked by human interaction. Her daily ritual was an act of self-protection as much as of self-expression. By carving out a discursive space, she was able to articulate her private thoughts and define her position in relation to others and to the world at large. Diary writers fashion a habit by choosing a medium; the creation of that mediated habit is always inspired by cultural conventions and prevailing fashions.[35] Quotidian acts such as diary writing should thus be regarded not only as stilled reflections but also as ways of constructing life. They always coexist amid a number of other communicative habits and culturally determined practices.

Cultural practices have become increasingly mediated in the past century: watching television, talking on the phone, taking pictures, and writing letters are only few of many potential communicative acts through which people articulate themselves. With the introduction of the Internet, some of these daily rituals are gradually changing, often fusing old practices with new conventions. For instance, the emergence of e-mail substantially transformed people's daily rituals of communication and interaction along with their sense of physical or psychological presence, just as the telephone changed communicative patterns along with notions of proximity and presence a hundred years earlier. For the contemporary blogger, the Internet is just one of a host of media through which to express agency, and blogging is one of many competing practices, such as speaking (both face-to-face and phone conversations), writing (letters, Short Message Services or SMS, e-mail), watching (television, film, photos), and listening (music, talk). The coexisting practices that fill the mediated lives of today's youngsters both complement and compete with each other; blogs offer several amenities that other media lack, such as the ability to combine extensive written comments with pictures, music, links, and clips, as well as the possibility to post something online to a large anonymous readership. Blogging is a combination of both oral and literate practices, such as diary writing, letter writing, reading, and conversation.[36] New hybrid rituals always emerge in connection to (and also in competition with) already existing practices, as they gradually create a new balance in the ecosystem of quotidian activities.

The networked computer is instrumental to the way in which bloggers simultaneously fashion their identity and create a sense of belonging. Blogging paradoxically complements and interferes with everyday live communication: blog entries are part of people's web of social circles through which they move and shape their lives. Some of these circles overlap; some do not. The generally reflexive nature of the lifelog has its place in the contact zones of everyday life, usually a mixture of real-life and virtual experiences. Although reciprocation is certainly not a condition for participating in the blogosphere, connecting and sharing are definitely written into the technological condition. Of all lifelogs present on the Internet today, some still resemble conventional paper diaries, whereas others have morphed into completely new interactive formats, firmly rooted in Internet culture.

Through their LiveJournals or Xangas, teenagers not only express themselves but also create a communal sense of values and thoughts deemed worthy of sharing. In a lifelog, one may blurt out confessions of loneliness and insecurity—behavior inhibited in face-to-face encounters—despite the fact that everyone in a peer group can potentially read these outbursts. In her journalistic ethnography, Nussbaum observes that bloggers usually don't talk about what they say online, even though in real life they may speak to each other on a daily basis.[37] Online posts may be read and responded to by immediate friends and relatives, and they may also invoke reciprocity from complete strangers, adding another dimension to the small world of immediate peers. Being able to choose the audience is a typical example of how technology and cultural practice interlock—the digital version of the lock-and-key diary. For example, the distribution features of LiveJournal allow users to decide with each posting to whom they make the content available—from "just myself" to "friends only" to "anyone." Defining one's readership is bound to define one's sense of inclusion in and exclusion from a community, whatever shape that community may take—actual or virtual, intellectually formative or emotionally supportive. In contrast to the paper diary, the weblog is part of a mediated continuum, a lived world in which the individual is always connected. However, just as the solitary basis of diaries turned out to be a myth, reciprocity is not a standard feature of blogs: still half of all Internet diaries are nonreciprocal.[38] Although the very medium that enacts blogging shifts the technological condition from isolation to connection, this does not mean that the cultural practices take on a new, pure default mode; on the contrary, old habits of diary writing coexist with new connected practices as they are gradually incorporated by a new medium.

The inclusion and exclusion of (potential) readers from one's lifelog constitutes an intricate game, the stakes of which are identity formation and community construction. Identity, as Australian media theorist Esther Milne claims, is always, in varying degrees, a performance: "It is the result of complex cultural, technological, economic and institutional forces rather than being a natural, somatic or psychological process that is fundamentally independent of historical influences."[39] Current complex forces are geared toward the swift and easy distribution of ideas. In the past, in order to expose oneself to a wider audience of unknown readers, a paper diarist was dependent on a publisher to print and distribute the diary, usually resulting in a considerable time lag between the moment of writing and of publication. But bloggers can make their own decisions concerning publication and distribution at the very moment of writing. Sharing intimate narratives with an anonymous readership is no longer a future possibility but an actual choice for webloggers; the effect of this technological option is immediacy—instant distribution, without intervention from a publishing institution. A survey held by the MIT Media Lab indicates that 76 percent of bloggers do not limit their readership in any way, and they have no idea who their readers are, apart from a core audience.[40]

Perhaps more than paper diaries, lifelogs foreground the intricate combination of technologies of self with technologies of affect: their prime function is to synchronize one's subjective experience with those of others, to test one's evaluations against the outside world. Besides being an act of self-disclosure, blogging is also a ritual of exchange: bloggers expect to be signaled and perhaps to be responded to. If not, why would they publish their musings on the Internet instead of letting them sit in their personal files?[41] Opening up one's secret diary to selected friends and relatives, and expecting them to do the same, is an old practice refurbished by bloggers. Attaching items of cultural contents is quite similar to swapping music albums, books, or personal accessories—a system of sharing symbolic meanings with friends that is firmly rooted in the material culture of gift exchange. But the potential to open up this process to an anonymous and potentially large readership is new; bloggers are constantly connected to the world at large, and they are aware of their exposure. By all means, the choice for privacy versus publicness, intimacy versus openness, appears to be a rhetorical or discursive effect, whether intentionally pursued, ignored, or simply forgotten.[42] As the above examples of lifelogs by AD patients already showed, intimacy and privacy as well as openness and publicness are less intrinsic features of the genre than implications of technological scripts and users' choices.

Based on their materiality and use, we often contend that whereas paper diaries are meant to fix experiences—freezing one's thoughts and ideas in time—blogs help users to synchronize their experiences with others'. Although there is a kernel of truth in this assumption, this distinctive function is all too easily ascribed to a material fixity of paper diaries as opposed to the evanescent quality of software or screen content. However, their ambiguous use as means of communication and storage is present in both diaries and lifelogs. Even though many consider blogging an ephemeral cultural practice—the equivalence of talking on the phone or sending short text messages—the desire for storage and retrieval is evident among users. For one thing, the fact that almost every software program contains an archive holding selected entries that go back to the very beginning of a person's blog signals a desire to build up a personal repository of memories. Although this hypothesis has never been empirically tested, it would be no surprise to find that bloggers, like teenagers using SMS or cell phones, value their lifelog's archival function as much as its communicative function.[43]

From the contemporary blogger's perspective, synchronizing experience and fixing experience in time are not at all contradictory functions, but they perfectly merge in today's lifelogs. Taking control of the evolving self, teenagers seem to take advantage of blogs to mold their living as well as their lived experience to reflect the kind of person they want to be perceived as being. Most youngsters favor the lifelog as a tool that lets them preserve their posted entries while concurrently allowing them to revise former entries.[44]

Lifelogs as Signifiers of a Cultural Transformation

Looking at the mental constructs, material technologies, and socio-cultural practices lifelogs engender, we can deduce an interesting reinvention of age-old rituals, newly attuned to the modalities of digitization. Like the writing of paper diaries, blogging is a process that helps shape subjective feelings and identity through affective connections, thus defining a sense of self in relation to others. Diaries and lifelogs are both acts and artifacts in which materiality and technology are contingent on their evolving use and users. Some seemingly conflicting genre features that have always existed converge in the face of hybrid practices, yet other paradoxes persist. Even though (networked) computers are gradually replacing pen and paper, multimedial materiality still reflects the kind of individuality formerly signified by handwriting and paper objects. The binary classification of diaries as strictly private documents and of lifelogs as public accounts that tends to inform the taxonomy of blogs in relation to diaries is equally fallacious. In the cultural practice of blogging, the need to synchronize personal experience easily blends with the desire to fix experience in time and to revise it over time. But however interesting lifelogs are in the perspective of their historical paper precursors, they are also momentous in their own right, signifying a techno-cultural transformation whose impact moves beyond its traces left on the World Wide Web. Tracking the evolution of blogging as a new hybrid practice, it is crucial to acknowledge how blogging strategies sustain old and construct new epistemologies, specifically paired-off notions such as privacy and publicness, intimacy and openness, and memory and experience.

As the examples of young people's use of blogs show, privacy and publicness are often assumed an inevitable part of the technological condition.

As pointed out above, though, privacy is a (deliberate or unintended) effect rather than an intrinsic feature of a blog's content, often achieved through one's (un)familiarity with the consequences of instant publication and global accessibility. Our norms and laws of privacy protection still rely on paper-based distinctions between ego-documents and public records; the boundaries that were often crossed in the past have become increasingly fuzzy for bloggers.[45] Obviously, there is no shame in sharing; bloggers take pride and find purpose in sharing. Instant publication, achieved by a simple click on the mouse, changes the rules of the game.[46] Especially youngsters' notions of privacy and publicness are riddled by contradictions: comments are personal yet readable by everyone. Old and new notions of privacy emerge alongside each other in the blogosphere as courts and lawyers are currently wrestling with emerging questions such as: Can entries posted with restricted access be stolen when they are posted on an open website? Are public officials or state employees free to speak their minds in the private sphere of restricted blog communities? It will take a number of years before this hybrid practice stabilizes and becomes grounded in social and legal norms.

For older bloggers with Alzheimer's disease, their struggle is less about issues of privacy versus publicness and more about intimacy versus openness. Like teenage bloggers, patients with dementia and AD find purpose in sharing their personal experiences as a means to counter the forgetfulness and loneliness forced upon them by the disease. But rather than striving for a discursive effect, these bloggers create affect in their attempts to open up their mental processes to readers, whether relatives or an anonymous audience. In more than one way, these patients manage to use the ambiguity inscribed in the technological and material condition of the lifelog to pair off intimate mental consciousness—a poignant sense of a changing brain impaired by increasing memory loss—with public awareness: the desire to inform and stay connected to a world that is increasingly erased from their consciousness. Partially serving as a therapeutic tool, the lifelog concurrently serves as a mnemonic device that perfectly mixes intimacy and candidness—terms that are no longer antonyms in the wired realities of patients' everyday lives.

Last but not least, lifelogs point at a profound change in contemporary notions of personal memory versus lived experience. The paper diary gave the erroneous impression that it was a petrified, unchangeable relic, stored in its authentic form and retrieved to invoke a past experience. When a diary's contents were published through an intermediate process of editing, printing, and distribution, we were mostly concerned with how the original words—presumably the recordings of experiences—matched the words published in print. The fusion of old and new technologies results in a hybrid tool that seamlessly combines editing and archival functions; blogging allows for preserving and revising entries at the same time. Blogging itself becomes a real-life experience, a construction of self that is mediated by tools for reflection and communication. In the life of bloggers, the medium is not the message but *the medium is the experience.* If the meaning of experience is slowly changing, so is the meaning of memory. As time proceeds, memories of experiences inevitably evolve; revising one's past inscriptions is a natural part of a process of personal growth. Rather than being fixed in objects, memory mutates through digital materiality. Although the Internet is often characterized as a transient, evanescent medium, lifelogs have both the ability to fix and the potential to morph.

It is exactly this hybrid function of the blog that helps Alzheimer's patients to sustain a sense of continuity in the face of harrowing mental changes in their self-consciousness. The lifelog becomes literally a memory device through which they can speak their minds, even as their states of mind appear increasingly confused in the eyes of relatives and loved ones. In the past, Alzheimer's patients were often recommended to look at old photographs to regain and retain an idea of past self, but in a way this meant a constant confrontation with the person they realized they would never be again, and for relatives these photographs carried the resentment of their "lost" loved one. Contrastingly, the lifelog is equipped to bespeak a changing self in the various stages of mental deterioration. Rather than lamenting the loss of a past persona—a memory to someone who has mentally if not yet physically disappeared from our lives—this enunciation at least refurbishes memory back into a real-life experience. Digital media, as amplifiers of affect, dramatically increase the rapidity and extent of inter-affectivity to reach global proportions, and this affect may in turn empower patients to keep themselves composed in their regular reports.

Bloggers are retooling the current practice of diary writing, meanwhile creating new types of mental scripts, cultural knowledge, and social interaction via their tools. The reciprocity inherent in networked systems points at a profound reorganization in social consciousness. Examining cognitive processes in relation to technological and material objects and to sociocultural practices, the analysis of concrete mediated memories may provide an index to understanding larger transformations—emerging hybrid practices that both reflect old conventions and construct new social norms. In a period of transition, concepts like privacy and publicness, intimacy and openness, and memory and experience continue to fluctuate. Contrasting conventional rituals to emerging practices, I try to unravel some of the complexities of change, even if they can never fully be grasped. As a teenager, my awkward attempt to share intimate growing pains with my mother was successful when she responded to me in kind, but my strategy backfired when my brother interceded in what I meant to be a targeted act of personal communication. Surfing on the Internet today, I am touched by mentally ailing parents who try to share their intimate mental and physical ordeal with their children, partners, and other loved ones, as well as with anonymous readers, and I can only hope they are responded to in kind. I am equally touched by botched attempts of teenagers to hide their intimate revelations in their Xangas and LiveJournals, and I suspect they will long remember this painful lesson in their future dealings with the machinations of privacy and publicity. The digital lifelog helps savor memories of a changing personality while also transforming notions of how we compose the self.

4

Record and Hold

In recent decades, recorded popular music has commonly been studied as either a vital component of people's personal memory or as a constitutive element in the construction of collective identity and cultural heritage. Psychologists and (neuro-)cognitive scientists have extensively researched the role music plays in the relation between emotion, individual identity, and autobiographical memory.[1] Sociologists, anthropologists, and cultural theorists, from entirely different academic perspectives, have studied recorded music as a significant part of our collective cultural memory and identity—a heritage that has grown over time and continues to evolve.[2] Engaging in shared listening, exchanging (recorded) songs, and talking about music create a sense of belonging, and relate a person's sense of self to a larger community and generation. In Chapter 1, I theorized a recursive connection between personal and collective cultural memory, and this chapter narrows this claim with regard to popular music.[3] People nourish emotional and tangible connections to songs before entrusting them to their personal (mental and material) reservoirs, but they also need to share musical preferences with others before songs become part of a collective repertoire that, in turn, provides new resources for personal engagement with recorded music. Naturally arising from the main contention of this book is the question of how personal memory and collective memory intermingle: how does recorded popular music stick to our individual minds and what makes it become part of a shared cultural memory?

Analysis of the interrelation between music and memory is based on the assumption, elaborated in Chapter 2, that the human's recovery of the past is simultaneously embodied, enabled and embedded. Autobiographical memories are embodied in the brains and minds of individual people, meaning that specific affects and emotions are attached to particular songs, a connection that is literally located in the body/mind because the human process of remembering is an "active, interpretative process of a conscious mind situated in the world."[4] Moreover, musical memories are enabled through instruments for listening. This was true before the advent of recording technologies, and still holds true in the age of digital reproduction. People become aware of their emotional and affective memories by means of technologies, and surprisingly often, the enabling apparatus becomes part of the recollecting experience. Songs or albums often are interpreted as a sign of their time not only because they emanate from a cultural-historical time frame, but also because they emerge from a socio-technological context.[5] Remembrance is also embedded, meaning that the larger interpersonal and cultural worlds stimulate memories of the past through frames generated in the present. Specific cultural frames for recollection, such as Internet forums or radio programs, do not simply invoke but actually help construct memory.

But how can we examine the embodied, enabled, and embedded nature of musical memories as a process and product of the human mind? This question poses an interesting challenge for neuroscientists and cognitive psychologists, but it requires the perspective of cultural theory to explore the connections between personal memory and collective heritage, between individual affect and social effect. Combining these perspectives, I propose to study the interrelation between personal and collective memories of popular music as they are constructed through stories of and about musical memory. Recollective experiences are often articulated in personal stories, and the analysis of narratives about music and memory are at the core of this chapter. To analyze the intertwining of personal and collective memories of recorded music, we turn to a readily available online set of narrative responses generated through a national radio event: the Dutch Top 2000. Every year, since 1999, a public radio station in The Netherlands (Radio 2) organizes a widely acclaimed five-day broadcast of the two thousand most popular songs of all times, a list entirely compiled by public radio listeners who send in their personal top-five favorite pop songs.[6] During the event, the station solicits comments from listeners—both aesthetic evaluations and memories attached to songs. Disc jockeys read those comments aloud during the live broadcast, and the comments are also posted in their entirety on a website. In addition, the station opens up a chatbox for exchanging comments. The result is an extensive database of comments and stories, opening an intriguing window into how recorded music serves as a vehicle for memory.

The first section analyzes these stories, posted in response to individual songs ranked in the listing, in terms of embodied affect and effect: how do individuals invest emotion and affect in recorded music? The second section examines how memory narratives often betray a distinct technological awareness: how did recording and listening technologies enable specific recollections? And finally, we turn to the embedded nature of the Top 2000: how do social practices, such as communal listening and exchanging recorded music, and cultural forms, such as the Top 2000, actually shape collective memories of the past? Remembering through music is more than just an individual act: we need public spaces to create a common musical heritage and identity.

Embodiment: Recorded Music, Memory, and Emotion

For a variety of reasons, we invest emotion, money, and time in compiling personal reservoirs of auditory culture. Like photographs or diary entries, music has a mnemonic function; listening to records helps inscribe and invoke specific events, emotions, or general moods.[7] Recorded music also has a formative function: young people, in particular, construct their identities while figuring out their musical taste. Building up a repertoire in one's memory (an inventory of familiar songs) and accumulating selected items of recorded music (a material collection of sound items) are considered important parts of one's coming of age. In Western countries, recorded pop songs are often signifiers of individually lived experiences; people select items of music to gauge their taste against those of others, and savor them in order to procure a sense of continuity. Albums or songs are items of culture that we select and collect, storing them into our minds or in our private "jukeboxes" and recalling them at a later moment in time. As American ethnomusicologist Tomas Turino convincingly argues, recorded music has the power to create emotional responses while also realizing personal and social identities.[8]

We often casually remark that certain music "gets stuck" in our minds and some songs "get under our skin." Once we start probing what is behind these clichés, we realize there is no easy answer to the enigma of how music gets nestled into our personal memories. Different parts and functions of the brain (cognitive, emotive, somatosensory) are involved in the remembrance of music. Repeated listening certainly helps; our cognitive memory can even be trained to retain melodic sequences for longer periods of time. As American cognitivist scholar Patrick Colm Hogan states: "The tendency of working memory to cyclic repetition combined with the exaggerated accessibility of a simple and frequently repeated tune gives rise to a situation in which the song is likely to cycle repeatedly through working memory. When this continues for a long time, we refer to it as 'having a song stuck in our head.'"[9]

The ability of recorded music to be replayed endlessly, repeating exactly the same performance, has aided the buildup of auditory memories in people's minds.[10] In addition to having a mnemonic function for the individual mind, repetition of music through media inscribes cultural experiences, literally playing them again and again. As media historian Lisa Gitelman describes in her concise history of the gramophone in America, "The phonograph introduced the intensity of true repetition to the performance of mass markets."[11] Being exposed to particular songs over and over again enhances their popularity, both in the private mind and in collective experience.

We cannot possibly retain all the music we hear in our lifetime, so there must be a mechanism accounting for why certain melodic rhythms get stuck in our long-term memory and others do not. In order to last, a song needs to catch our attention, somehow standing out from other experiences or perceptions.[12] In explaining why music gets under the skin, cognitive scientists often refer to somatosensory reflexes and emotions as key factors in memory formation. We commonly think of intuitive responses to music as articulations of taste, but besides rationalized judgment, aesthetic pleasure or dislike also result from simple emotional arousals. A manifestation of our core consciousness, the perceptions of sounds may elicit direct physiological responses, such as shudders or body hair standing on end. In explicit accounts of musical reminiscing, such as the comments posted to the Top 2000, people often describe such physical reactions. Many of these comments are brief, expressing a simple emotion or aesthetic judgment ("this

song makes me happy" or "the tonal arrangement of this song is sublime").
Bearing witness to music's emotive and somatosensory reflexes are the numerous comments inferring a listener's somatic response when exposed to
particular music: "this song sends shivers down my spine" or "each time I
hear this song, I get goose flesh." Visceral reactions abound in the thousands of comments posted to the Top 2000 website; however, they are the
least interesting as narratives.

These reflexes, however significant in explaining emotional like or
dislike, cannot satisfactorily account for how music gains a permanent
presence in our autobiographical memories.[13] In order to find a more
agreeable explanation, I propose two complementary views. The first—
neurocognitive theory detailing the brain's and the mind's involvement in
constructing personal memory—is touched upon only briefly, because the
second explanation, stemming from a cultural-semiotic perspective, expounds on the first. As elucidated by Antonio Damasio, emotions differ
from feelings, in a sense that feelings only occur after we become aware in
our brain of emotional arousals.[14] These feelings are then inscribed as
mental image maps, maps that do not remain the same but mutate with
each recall; moreover, during acts of reminiscence, remembrance and projection coil beyond differentiation. Based upon Damasio's conjecture, we
can assume that our memory of a song is more durable when we affectively
invest in making it stick—that is, by constructing meaning for and around
a musical item. But how do songs make sense to us in the long run? Damasio answers this question by explaining autobiographical memory as a
function of extended consciousness that involves emotions and feelings. In
addition to storing the sound of an object we hear, our memories also retain emotional reactions to it, as well as our mental and physical state at
the time of apprehending. Out of that sensation or feeling, we create a
(non-language) map or image of this event in our core consciousness, a
story that also becomes verbally present in our minds by the time we focus
on it. Upon later recall, recorded songs work as triggers, bringing back
waves of emotion tied specifically to a time, an event, or a relationship or
evoking more general feelings. This wordless storytelling precedes language and happens entirely inside the brain; memory for recorded songs
appears to hold longer when people turn emotion-infested sounds into internal narratives. Damasio even speculates that the mechanism accounts
for why and how we end up creating drama: verbal stories, books, televi-

sion plays, and so forth may directly derive from the wordless stories first created in our minds.[15]

Damasio's approach highlighting internal narratives is surprisingly complementary to cultural-semiotic theory relating musical memory to individual and social identity. Thomas Turino has attempted to counter the intriguing problem of musical memory by approaching music as a system of signs; he uses Charles S. Peirce's notion of indexicality to explain how music is not about feelings but rather involves signs of feeling and experience.[16] Musical signs, he says, are sonic events that create effects and affects in a perceiver the way a falling tree creates waves through the air. Rational effects or conscious responses are responses that involve reasoning: the interpretation or appreciation of music. The "secondness" or affect of music lies not in the sounds or words per se, but in the emotions, feelings, and experiences attached to hearing a particular song. Musical signs thus carry strong personal connotations betraying an emotional investment; at the same time, however, members of a social group share indices proportional to common experiences.[17] In sum, musical signs integrate affective and identity-forming meanings in a direct manner, and it is through the recollection of songs that we may come to see the nature of this integration.

Damasio's neurocognitive speculation and Turino's cultural semiotic conjecture cannot be emprically tested, but applying narrative analysis to stories that relate how people feel affected by recorded popular music offers an insight into the connection of personal and collective memory. Many respondents to the Top 2000 create images or stories around certain songs to help them communicate a particular feeling or mood or to express their affective ties to particular songs. Through these stories, we learn how people came to invest emotionally in a song, how the song came to mean anything to them in the first place, and how they retained that attached meaning—a meaning they like to share with a large, anonymous audience. As Damasio predicts recall includes the experience of listening and the emotive state at the time of apprehending. Compare, for instance the first reaction to John Lennon's song "Imagine" with the next reaction, posted in response to the U2 song "With or Without You":

It was 1971, I was waiting on a boat someplace in Norway when I heard this song for the first time. It was such a perfect day, everything was right: the weather, the blue sky, the peaceful tidal waves in the fjord matching the melodious waves of music. There are moments in life that you feel thoroughly, profoundly happy. This was such moment, believe me. (posted by Jan from Eindhoven)

My father died suddenly in November of 1986. That night we all stayed awake. I isolated myself from my family by putting on the headphones and listening to this song. The intense sorrow I felt that night was expressed in Bono's intense screams. I will never forget this experience, and each time I hear this song I get tears in my eyes. (posted by Jelle van Netten from Woudsend)

Memories tied in with extreme pleasure or intense sorrow, like the ones above, are likely to stick to our minds, due the brain's tendency to store sound perceptions along with their affects and moto-senseo impact. The (explicit) narratives created out of these memories are deemed worthy of sharing because they exemplify both universal and intimate experiences.

But memories are not always tied in with specific affective experiences; they may also evoke the mood of a time, place, or event in which the song first became meaningful. Some songs trigger a more general mood or atmospheric sensation—an almost Proustian invocation of the past. The Beatles recording of "Penny Lane" seems especially conducive to such invocations of sensory moods:

When this song first came out, I was three years old. Every time I now hear this record I can see a picture of my grandmother's living room, because I lived there at the time. This is very odd, because I can't remember anything else from that time, and this record puts me back into that time and place. It is my very first musical remembrance. (posted by Anja from Rosmalen)

In this comment, a general longing for the mood of a past era is associated with lived experience, even though the experience is somewhat blurred and sensuous. Mental maps, as neuroscientific theories show, are derived partly from the object itself and partly from the auditory, visual, olfactory, or other perceptions triggered in our minds.[18] Respondents also frequently report smells and tastes to be invoked by familiar songs.

The comment above implicitly suggests that the memory aroused upon hearing the song is an exact repeat of the original listening experience. The idea of a recording reiterating its exact same content each time it is played is subconsciously transposed to the experience attached to hearing the music. People's expectation to feel the same response each time a song is played stems from a craving to relive the past as it was—as if the past were a recording. Many of us want our memory of the original listening experience to be untainted by time, age, or life's emotional toll. And yet, it is improbable that repeated listening over a lifespan would leave the original emotion (if there ever was such thing) intact. Instead, the original listening experience may be substituted by a fixed pattern of associations, a pattern that is likely to become more brightly and intensely colored over the years.[19] Memories change each time they are recalled, and their contents are determined more by the present than by the past. As Geoffrey O'Brien eloquently puts it in his musical memoir: "The age of recording is necessarily an age of nostalgia—when was the past so hauntingly accessible?—but its bitterest insight is the incapacity of even the most perfectly captured sound to restore the moment of its first inscribing. That world is no longer there."[20]

Cognitive research confirms that musical remembrance alters with age. An American clinical study shows a significant difference between how older and younger people remember recorded music; whereas young adults tie in recorded music with memories of specific autobiographical events, seniors use familiar songs as stimuli to summon more general memories and moods from the past.[21] Recorded music infested with feelings elicits stronger—even if less specific—autobiographical memories later on in life. Because the narratives posted to the Top 2000 website do not systematically disclose the respondents' ages, I cannot use them to confirm or disprove the researchers' empirical observation. In general, though, respondents who give clues to their age as being over forty-five tend to refer more to nostalgic moods triggered by specific songs than respondents who identify themselves as being young adults. However, this might just as well be attributed to the fact that a majority of songs featured in this collective ranking were popular in the era when baby boomers came of age.

The observation of neuroscientists and cognitive psychologists that musical memory is an emotional investment has been corroborated by both sociologists and cultural theorists. Australian cultural geographers John Connell and Chris Gibson argue, for instance, that "music can evoke memories of youth and act as a reminder of earlier freedoms, attitudes, events; its emotive power . . . serves to intensify feelings of nostalgia, regret or reminiscence."[22] And yet, it is too simple to assume that music triggers memories; recorded music may also construct a cognitive framework through which (collectively) constructed meanings are transposed onto individual

memory, resulting in an intricate mixture of recall and imagination, of recollections intermingled with extrapolations and myth. One listener, in response to the same Beatles song mentioned earlier, comments on the oddity
of certain music invoking a historical time frame she did not live through:

This song elicits the ultimate Sunday-afternoon feeling, a feeling I associate with
cigarette smoke, croquette (the Dutch variant of the hamburger, jvd) and amateur
soccer games. This feeling marks my life between the ages of five and fifteen. A
nostalgic longing of sorts, although I have to admit I was not even conceived
when this record became a hit song. (posted by M. Klink from Leiden)

The respondent transposes the general mood of an era onto her
childhood, even though these periods are distinctly apart. It is not un-
thinkable that she has projected a general impression of the zeitgeist onto
this particular song.[23] Recall and projections thus curl into one story, even
when the respondent realizes that the memory is not rooted in actual lived
experience. Mixing memory with desire or projection is a common phenomenon acknowledged by cognitive scientists and neurobiologists.[24]

At this point, the unmistakable intertwining of personal and collective memory, as theorized by Turino, is obvious: narratives about music
often braid private reminiscences into those of others or connect them to
larger legacies. Certain songs become "our songs," as they are attached to
the experience of various collectives, whether families or peer groups. Verbal narratives appear to be important in the transmission of both musical
preferences and the feelings associated with them to the extent that it becomes difficult to distinguish lived memories from the stories told by parents or siblings.[25] This does not mean, of course, that children uncritically
adopt their parents' memories or, for that matter, their musical taste, but
young people construct their own favored repertoire by relating to peers as
well as to older generations, either positively or negatively. Musical memories become the input—resources to adapt or resist—in the edifice of one's
own repertoire. A few assorted comments posted to the Top 2000 website
may illustrate this. One respondent, reacting to the Doors hit song "Riders
on the Storm," writes:

One of the things my father passed on to me was his musical taste. His absolute fa-
vorite was Jim Morrison and as a child, I would sing along with every Doors' song.
Remarkably, my father thought "Riders on the Storm" to be one of the worst
Doors songs, but I think it's one of their best. (posted by Joanna from Heerlen)

And another listener attributes her fondness for the pop song "You're So Vain" by Carly Simon to fetal exposure:

When my mother was pregnant with me, in 1973, my father bought her this album as a present. They played the record innumerable times. As long as I am aware, this has been my favorite pop song, but I only found out about my parents' story several years ago. Who knows, listening to music in the womb may have an effect on a person's musical taste! (posted by Harriette Hofstede from Dordrecht)

Musical memories can thus be understood as an intergenerational transfer of personal and collective heritage, not only by sharing music but also by sharing stories; many comments posted on the website testify to this circular cultural process. Like photographs, recorded songs relate personal memories. That older people are eager to pass on their stories along with their preference for certain recorded music is therefore not surprising.

Damasio's conjecture about the mind's involvement in autobiographical recall, suggests that narrated audio impressions help glue recorded music to people's cultural memory. Understood in terms of bodily affect, the mind is a sewing machine that quilts personal memory onto recorded music, stitched together by emotion and feelings. Whether tied in with specific experiences or general moods, stories appear a distinct aid in remembering the mental associations attached to a particular kind of music. These stories, like the memories themselves, are likely to change with age, and as much as we like to capture the original affection triggered by music, we want the story to hold that feeling for future recollection. Yet stories, like records, are mere resources in the process of reminiscence, a process that often involves imagination as much as retention. In other words, our personal musical repertoire is a living memory that stimulates narrative engagement from the first time we hear a song up to each time we replay it at later stages in life. It is this vivid process of narrative recall that gives meaning to an album and assigns personal and cultural value to a song.

Enabling Technologies: Recorded Music and "Techno-stalgia"

Technologies and objects of recorded music are an intrinsic part of the act of reminiscence; even though their materiality alters with time, often generating resentment, their aging may partially account for our very attachment to these objects. Personal memory evolves through our interactions with these apparatuses (such as record players, compact disc players, radios) and material things (such as records, cassettes, digital files), as both are agents in the process of remembering. Media technologies and objects are often deployed as metaphors, expressing a cultural desire for personal memory to function like an archive or a storage facility for lived experience. When it comes to music, it is easy to see where this metaphorical notion originated. The record's presumed ability to register—to record and hold—a particular mood, experience, or emotional response can be traced back to the its historically ascribed function as a material-mechanical inscription of a single musical performance. It is almost a truism to expect recording equipment to replay the presumed original sound of a song, notwithstanding our awareness that objects and apparatuses—like bodies—wear out, age, and thus change over time. The "thingness" of recorded music is unstable and yet this knowledge does not prevent a peculiar yearning for the re-creation of audio quality as it was first perceived, evidenced for instance by the recent vinyl nostalgia accompanying the surge in compact disc sales.[26] People who use recorded music as a vehicle for memories often yearn for more than mere retro appeal: they want these apparatuses to reenact their cherished, often magical experiences of listening.

It may be illustrative to filter this kind of techno-stalgia from the comments posted on the Top 2000 website, espousing the integrality of technology to people's reminiscing. Many respondents recall the sound equipment through which they first heard a particular song, emphasizing how it defined their listening experience. Writing in reaction to the Beatles song "The Long and Winding Road" a woman writes:

The first time I heard this song I almost snuggled into my transistor radio. This was the most beautiful thing I had ever heard. When I got the Beatles' album, I remember pushing the little Lenco-speakers against my ear (they were sort of the precursor of the walkman). Whenever this record is played again, I get on my knees, direct my ears downward, pushing them toward the speakers on the floor. I still want to live in this song. (posted by Karin de Groot from Rotterdam)

In this comment, the experience of listening seems inextricably intertwined with the (primitive) equipment that enabled its broadcast—and that memory has become partial to its reenactment in contemporary stereo systems. Needless to say, the reenactment never brings back the equipment and context of the original sound—a fact the respondent is very aware of—but certainly brings about the intended affect.[27]

In other instances of reminiscence, the role of technology should be understood indexically rather than metaphorically—adhering to Thomas Turino's Peircian apparatus—as it stands for taking control over one's sonic space. Memories of the original listening experience often include allusions to the newly acquired freedom to listen to these songs, alone or with friends, outside the living room where the soundscape was usually controlled by the musical taste of parents. The 1960s ushered in a period of "private mobilization," to borrow Raymond Williams's renowned term, so it is not surprising to find that many respondents, recalling impressions from that era, take notice of sound technology's bearing upon their coming of age.[28] Many are still committed to the sounds mounting from the radio (especially transistors and car radios), a medium that first confronted the baby-boom generation with pop music. In contrast to music played on personal stereos (record players or tape recorders), radio sound is ephemeral. Listening to music on the radio often allows for a momentary inner sensation that the listener is part of something larger; it creates relationships between self and others that contribute to an individual's sociality.[29] Narratives that testify to the liberating role of music technology abound on the website postings to the Top 2000. Read, for instance, this reaction to the Herman Hermits song "No Milk Today":

Because this was the first song to wake me up to the phenomenon of pop music in the years 1966–1970, it reminds me of how magical it felt to just listen to my small transistor radio, often secretively, because I needed to hide it away from my parents. When I listen to this song now, I turn up the sound as much as I can, preferably when I am driving my car and listening to old tapes. (posted by Maarten Storm from Leusden)

For this respondent, hi-fi equipment was (and remains) a technology that endowed him with the liberty to create his own sonic space. There are many responses similar to this one, all attesting to the importance of stereos in forming an autonomous sense of self and the mental-physical room to develop one's personal musical taste. Some respondents explicitly relate how their attempts to capture favorite songs played on the radio resulted in tapes of very poor quality, and yet they still treasure their amateur recordings not in spite of, but because of their obvious technical shortcomings.

The eminent awareness that objects and technologies are subject to inevitable erosion underscores their quality not only as objects but also as agents of autobiographical memory. Many respondents to the Top 2000 website remember the song in the gestalt of an object they once bought. For instance:

I was eleven years old when Paul Simon's "Kodachrome" become a hit. I liked this song immediately and purchased the single with my hard-earned savings. Even after thirty years, I still cherish this object. (posted by Liesbeth)

Objects also become agents once people start to rerecord music—taping songs in order to literally appropriate them and use them for endless replay. Fiddling with lo-fi tape recorders to catch radio broadcasts, or playing vinyl records over and over again—even as their quality deteriorates as a consequence of multiple use—somehow contributes to the intensity of recorded music stored in memory. As one anonymous respondent admits on the Top 2000 website: "This album finally collapsed on my record player, completely worn out by its relentless owner. I tried to obtain a vinyl replacement, but was unsuccessful."

Audio artifacts and technologies apparently invoke a cultural nostalgia typical for a specific time and age. The ability of digital recording techniques to meticulously recapture a worn-out recording and reproduce its exact poor auditory quality may offer only partial solace to a cultural yearning. Joseph Auner suggests that every new medium in a way authenticates the old, meaning that each time a new audio technology emerges on the scene, the older ones become treasured as the authentic means of reproduction or as part of the original listening experience.[30] In the digital era, scratches, ticks, or noise can be removed from tapes to make old recordings sound pristine, but they can also be added to make a pristine recording sound old. Sound technologies thus figure in a dialogue between generations of users: think, for instance, of a young musician's sampling of original pop songs into digital sound experiences or the creative use of old telephone sounds as ring tones on teenagers' cell phones.[31] The dialogue with outdated technologies, frequently used in contemporary pop songs, symbolizes recorded music's ineffable historicity. Paradoxically, sound technologies are concurrently agents of change and of preservation; with the new digital technologies, sonic experiences of the past can be preserved and reconstructed in the future.

Incontrovertibly, the materiality of recorded music influences the process of remembering. The term "recorded music" has become the rather generic container for vinyl albums, cassette tapes, compact discs, and MP3 files stored in computers. But the status of these items varies and that variation affects their function in memory formation. Each song listened to from live radio, records, cassette tapes, or MP3 players has a different emotion attached to it. Prerecorded CDs and records are more valuable as objects to hold on to and collect, whereas MP3s and cassette tapes have a different function: they are more often a backup or backlog. As music theorist Mark Katz shows in his study on sound capturing, which includes a survey of young downloaders of recorded music, a large majority of respondents still buys prerecorded CDs, often after having listened to them in rerecorded form or after having shared them in whatever mechanical or digital form: "The tangibility of the CD is part of its charm. A collection is meant to be displayed, and has a visual impact that confers a degree of expertise on its owner."[32] In semiotic terms, the indexical function of the musical sign is bound up with its auditory materiality: hearing a familiar song on the radio constitutes a different memory experience than playing that very song from one's own collection, perhaps even more so when these recordings are played from MP3 formats. As one respondent to the Top 2000 puts it:

It is so strange: I keep most songs [featured in the Top 2000] on CDs and I have the entire list of songs stored in MP3 format on the hard disk of my PC, so I can listen to these songs any day any time. And yet, I only swing and sing along with my favorite songs when I hear them on the radio, during this yearly end-of-the-year broadcast. (posted by Jaap Timmer from Winterswijk)

The transfer of emotive affection from the brain onto the technology and materiality of audio recordings shows how memory acts out in the spaces between individual reminiscences and shared experiences. These same narratives disclose how materiality and technology often become integral to memory, something that is unlikely to change with the advent of digital equipment. As long as listening to music remains a mediated experience, memory will be enabled and constructed through its material constituents.

Embedded Memory: Shared Listening and Exchange

From our explorations of the embodied nature and enabling aspects of recorded music with regard to memory, it is clear that individual and collective memory are inextricably intertwined. Memories attached to songs are hardly individual responses per se; recorded music is perceived and evaluated through collective frameworks for listening and appreciation. Individual memories almost invariably arise in the context of social practices, such as music exchange and communal listening, and of cultural forms, such as popular radio programs, hit lists, music programs, and so on. These social practices and cultural forms appear almost inseparable from the memory of actual songs; as a sign of their time, popular songs create a context for reminiscence.

Through these practices and forms individual memories become collective vehicles for identity construction. Sociologist Tia de Nora observes in her ethnographic study of young adults and the way they use recorded music in everyday life how audio equipment implicates individuals evolving into social agents.[33] Throughout their entire lives, people build up mental and material reservoirs of musical preferences. Selected songs may become meaningful when they are consigned aesthetic value, when they are associated with lived and shared experience, or when linked up with spatial configurations. Since the introduction of sound recording in the last decade of the nineteenth century, sonic experiences have been assigned meaning as collective memories through performative rites, like shared listening and exchanging music.[34] Listening to recorded music has always been a social activity: listening with peers or sharing musical evaluations with friends helps individuals to shape their taste while concurrently constructing a group identity. It is therefore understandable that the sociability of listening to pop music becomes an inherent part of people's memories. For instance, one respondent adds the following general comment to the Top 2000 chatbox:

It was 1976, and with a number of friends I organized disco events for the local soccer club. These events always turned into choose-your-favorite-pop-song tournaments. The Top 2000 reminds me of these days. (posted by Henk Vink)

A large number of comments posted to the Top 2000 website relate how groups of people—varying from three-generation families at home to labor crews and office personnel—stay tuned to the nonstop five-day event and listen as a group. One woman confides to the chatbox how listening to the Top 2000 during a house remodeling project facilitates previously deadlocked communication between a grandfather, parents, and children.

The radio event engenders collectivity at the same time and by the same means that it generates collective memories; the actual sharing of music and singing along with a group hence becomes part of the emotion triggered by a song.

Some sound technologies, by nature of their hardware, promote listening to music as a solo activity, but can still be deployed in social activities. Ever since the emergence of the Walkman in the 1970s, personal stereos have been associated with the construction of individual sonic space. As British sociologist Michael Bull argues, personal stereos can function as a form of "auditory mnemonic" in which users attempt to reconstruct a sense of narrative within urban spaces that have in themselves no narrative sense to them.[35] And while it is true that the Walkman—and more recently the MP3 player and Discman—are designed with individual urban listeners in mind, these recording technologies can also be put to social use and serve as collective listening instruments. As we can read in the posting of an eighteen-year-old respondent to the Top 2000 website, who commends the 1961 song *"Non, je ne regrette rien"* performed by Edith Piaf:

Last summer, half a dozen of my classmates drove to France to celebrate our high school graduation. We played a lot of oldies, and as both cars had their own iPods attached to the stereo system, we sang along as loud as we could with our self-compiled repertoire. Now we've all gone off to different colleges, but next month we'll have a reunion and I'm sure we'll bring our iPods along, so we can bring back some cherished memories. (posted by Willem van Oostrum from Utrecht)

The rather novel act of plugging the iPod into the car's stereo system, allowing the youths to collectively listen to and sing along with the songs stored on the device, is inscribed in the narrative recollection of a generation of young adults; they consciously create their own sonic memories, using the newest devices to re-create golden oldies. The MP3 player, rather than being a mere vehicle for individual listening and storage of favorite songs, thus figures as agency in the conscious process of building up a collective memory.

Besides collective listening, remixing and exchanging songs are important means for constructing a collective platform for shared memories. Memory and identity are ineffably bound to us in the way we re-create given formats; interventions in prerecorded cultural forms propelled by the music industry are more than symbols of individual appropriation.

Through rerecording, mixing, and remixing ready-made formats (such as albums), people invent their own memory products—they use the available templates as input for creating idiosyncratic compilations. Compiling one's own favorite collections, to give away or share, has become a popular social practice since the emergence of tape recorders. Mixtapes typically have a strong emotional and personal touch to them. They are anchors of personal memory; each time they are replayed, they rouse the expectation not of relived experience per se but of a familiar, anticipated order determined by its maker. That order of songs is reified on tape and in memory as a coherent, customized unit of musical replay. Good and lasting mixtapes become personalized albums: unique recordings made to stick to one's mind and reflect the compiler's taste. If any feature stands out from the plethora of mix-and-burn software products currently available on the Internet, it is the capacity to compose musical collages to generate, incite, or control certain moods—feelings, occasions, emotions, or frames of mind.[36] Self-compiled mixes provide a contextual narrative for channeling concrete feelings or experiences; they "burn" certain impressions into the mind and thus confine audio-image maps to memory.[37]

Idiosyncratic compilations are commonly not restricted to private use but are made to share with friends or, increasingly, with unknown recipients via playlists and home-burned CDs. MP3 files lend themselves particularly well to multiple and effortless exchange, although this digital materiality, in recent years, has become the center of a controversy over the legality of freely downloading recorded music. As Jonathan Sterne argues, though, the new material quality of recorded music obviously deserves to be examined in its own right as it generates new cultural practices involving mixing and sharing.[38] Digital mixes of songs copied onto compact discs or playlists, made for sharing or distributing on the Internet, are both continuations of and variations on earlier auditory exchange rites.[39] The mix-and-burn culture, favoring the reconfiguration of digital songs into playful aggregations, signifies an individual's desire to contribute to the formation of communal tastes and group identity; with the advent of CD-burning software, mixing and burning compilations has become technically easy, so the social practices of sharing music and (symbolic) gift exchange are increasingly part of the apparatus's script. Many comments in the Top 2000 chatbox testify to listeners' inclination toward audio-creativity, for instance by offering one's own compilation of live recordings of the songs played on the radio. The chatbox at times functions as a platform for the exchange of homemade selections, thus becoming a venue for sharing creative re-collections—a way of embedding one's idiosyncratic choice in a community of listeners.

To sum up, cultural practices like communal listening, mixing, and swapping recorded music appear crucial in understanding how and why we construct shared memories through embedded experiences: musical memories are shaped through social practices and cultural forms as much as through individual emotions. New digital technologies allow music fans to customize their favorite songs and use them as symbolic resources in the construction of collective identity and community. But let us now turn to the role the Top 2000, as a cultural form, plays in inculcating a sense of collectivity in its listeners. Should we look upon this event the way we regard rankings of pop music and radio programs? In what sense do the Top 2000 stories of individual reminiscence contribute to a sense of collective memory and communal cultural heritage?

The Top 2000 as Collective Cultural Memory

The Dutch Top 2000 nicely illustrates how mediated memories are shaped precisely at the intersections of personal and collective memory. Mental mappings, sound technologies, and sociocultural practices constitute the channels for shaping individuality while concurrently defining the larger collectivity we (want to) belong to—ensuring autobiographical as well as historical continuity. Through embodied affection, enabling technologies, and cultural embedding, recorded music becomes part of our collective memory at the same time and by the same means as it gets settled into our personal memories. Viewed from a neurocognitive angle, we often engage with recorded music by stitching emotion or lived experience onto musical impressions, hence conjuring up mental maps—internal stories that are later recalled as part of our musical memory. Theorized from a semiotic-cultural perspective, personal emotions and affects attached to songs are articulated in explicit memory narratives that people like to exchange—reminiscences of lived experience expressed through musical preferences. These stories not only are about emotions triggered by music, but they also directly bespeak musical memory as it relates to personal and group identity, not seldom handed down generation to generation.

Through collective experiences, embedded in social practices and cultural forms (shared listening, rerecordings, the exchange of music compilations), people build up collective reservoirs of recorded music that stick to the mind and, in terms of collectivity, become our cultural heritage.

Building a national heritage of favorite popular music is obviously the propagated goal of the Top 2000, and of course, an important key to its success. The eminent value of creating collective musical repertoires, as American historian William Kenney points out, has proved vital to the "ongoing process of individual and group recognition in which images of the past and present could be mixed in an apparently timeless suspension that often seemed to defy the relentless corrosion of historical change."[40] The Dutch Top 2000 constructs and reflects a national consensus about which songs in this particular moment in history constitute the people's national heritage. Even if only 15 percent of the elected pop songs are of Dutch origin and/or are performed in a language other than English, the selection is a quintessential national event. How important compiling this list is to the formation and (re)confirmation of Dutch identity appears from the many comments posted by expats and emigrants tapping into the event from all over the world. Without exception, they praise Radio 2's initiative to make this five-day event available through broadband Internet. As one respondent residing in Australia claims, "The Top 2000 enables you to travel to the homeland of your youth, going home without leaving home." The event produces a collective, national identity because the memories invoked are themselves the result of a particular kinship between listeners and nation.[41]

But as important as creating a cultural heritage may be as a key to understanding the Top 2000's popularity as a national event—more than half the population of the Netherlands plugs into the yearly event—its success can hardly be explained by the nation's craving for a collective repertoire. We cannot overstate the significance of this event as a platform for exchanging personal stories of musical reminiscence—a crucial function in the formation of collective memory. Of course we can never speak of a unified collective memory; instead, there are numerous networks, platforms, and sites for constructing versions of a communal past. Collective memories are achieved through negotiation and consensus building among a variety of remembering subjects. It is in the public spaces between individuals, technologies, and communities that memory is shaped and negotiated. The process of narrating, discussing, and negotiating musical heritage and personal versus collective identity is far more important than the ultimate result; interactive participation of listeners is the goal, not the means by which a ranking is compiled.

One might argue that the collective nature of the Top 2000 can just as well be explained by its trendy catering to a participatory, (inter)active audience; the event, after all, fits in well with the current boom of audience participation contest shows on radio and television. However, there are several arguments to rebut this explanation. For one thing, it is not a coincidence that the event is staged through public rather than commercial radio. Staging the Top 2000 is unlike oldies listings by commercial stations catering to a retro experience. As stated earlier, many if not all of the songs featured in the Top 2000 can be heard by tuning into one of the many commercial oldies stations abounding on the airwaves, and they can be listened to on demand by downloading or pulling them from one's private collection. Cultural historian Paul Grainge proposes the relevant distinction between nostalgia as a commercial mode and a collective mood. In the example of the Top 2000, nostalgia emanates from a collectively experienced mood, in contrast to a conception of nostalgia as a consumable stylistic mode espoused by commercial outlets such as the Top 40 or oldies stations.[42] I concur with Grainge that the oldies station phenomenon can be better explained by the industry's imperative to find profitable market segments and niche consumption than by a presumed generational longing for an idealized past. Grainge's concept of nostalgia as an experienced mood links up with my definition of collective cultural memory: it connects personal affect and emotion to collective identity and heritage via recorded music.

In addition, I think the participatory, interactive nature of the Top 2000 is geared toward discussion and exchange rather than commercial call-in activities and revenue-generating voting strategies. Through a combination of a radio event, website, and television broadcast, this multimedia platform offers space for consensus building on a national heritage of pop songs, and it simultaneously serves as a podium for collective nostalgia and communal reminiscence. Audience participation is of course a constitutive element of the ultimate ranking list—indeed, without individuals sending in their favorite top fives there would be no Top 2000. And without the hundreds of thousands of comments filling the chatbox and participants sharing stories through web postings, there would be no event. It is rather the desire to couple personal memories onto collective experience and the need for a platform for the exchange of musical memories that constitutes the repeated success and public impact of this event.

The extensive archive of responses generated by the Dutch Top 2000 constitutes an interesting source of data on personal musical memory and cultural heritage formation. It opens up new perspectives on the importance of public space for sharing personal stories and constructing a collective musical kinship, which in turn feeds our individual creativity and identity. The commercial domain, although an important provider of resources for a common culture, seems to have less and less tolerance for the necessity of building up collective reservoirs, public involvement, and creative exchange. Virtually void of commercial push-and-pull mechanisms, the Top 2000 offers space for narrative engagement and gives room to what American legal scholar Lawrence Lessig calls a "creative commons."[43] Whereas (digital) culture, according to Lessig, opens up ample opportunities for strengthening the public domain and for promoting individual creativity, the commercial music industry often impedes exchange and participation.[44] The Top 2000 encourages both individual memories and collective reminiscence; attaching emotion to recorded music is not only imperative to the formation of personal identity, but it is instrumental in imagining collectivity. Indeed, the public domain, or the creative commons, is vital to keep alive a vibrant heritage of old and new music, because it provides individuals with cultural resources to understand their own pasts and guarantee a shared interest in a communal future—both essential forces in people's long-term commitment to music. By ignoring insights in how recorded music functions as cultural memory—the linchpin between individual remembrance and collective cultural heritage—we may deprive ourselves of an important enticement for future growth.

5

Pictures of Life, Living Pictures

A student recently told me about an interesting experience. She and four friends had been hanging out in her dormitory room, telling jokes and poking fun. Her roommate had taken the student's camera phone to take a group picture of the friends lying in various relaxed positions on the couch. That same evening, the student had posted the picture on her photoblog—a blog she regularly updates to keep friends and family informed about her daily life in college. The next day, she received an e-mail from her roommate; upon opening the attached JPG file she found the same picture of herself and her friends on the couch, but now they were portrayed with dozens of empty beer cans and wine bottles piled up on the coffee table in front of them. Her dismay at this unauthorized act of photoshopping only intensified when she noticed the doctored picture was e-mailed to a long list of peers, including some people she had never met or only vaguely knew. When she confronted her roommate with the potential consequences of her action, they engaged in a heated discussion about the innocence of manipulating pictures ("everybody will see this is a joke") versus the incriminating potential of photographs ("not everyone may recognize the manipulation") and the effect of their distribution, which might have a less transitory impact than anticipated ("where do you think these pictures may show up?").

In recent years, the role and function of digital photography seem to have changed substantially. In the analog age, personal photography was first and foremost a means for autobiographical remembering, and photographs usually ended up as keepsakes in someone's album or shoebox.[1] They were typically regarded to be a person's most reliable aid for recall and for verifying life as it was, despite the fact that imagination and projection are inextricably bound up in the process of remembering.[2] Photography's functions as a tool for identity formation and as a means for communication were duly acknowledged, but they were always rated secondary to its prime purpose.[3] The recent explosion in the use of digital cameras—including cameras integrated in other communication devices—may be a reason to reconsider photography's primacy as a tool for remembering. Deployment of the digital camera is now quite commonplace, but it is a truism to say that technology has impacted the way we "picture our lives." This chapter explores how digital photography, in conjunction to a changing cognitive mindset and sociocultural transformations, is reshaping personal cultural memory.

As the student's anecdote illustrates, digitization is often considered the culprit of photography's growing unreliability as a tool for remembrance; but in fact, history shows the camera has never been a dependable aid for storing memories, and photographs commonly have been twitched and tweaked in the process of recollection. The story above raises several intriguing questions concerning the revamped role of personal photography in contemporary digital and networked culture. First, there is the question of (digital) manipulation and cognitive editing: what is the power of digital tools in sculpting autobiographical memory and forming identity? How do we gauge new features that help us edit our pictures and make our memories picture perfect? Besides seeing photography as an extension of mental processes, we need to account for its materiality and performativity. Pixeled pictures on a computer screen have a different touch and feel to them than their laminated precursors; photoblogs are not the equivalent of digital photo albums, as they elicit distinctly new presentational habits. And finally, we need to ask how these changes evolve along with sociocultural practices: photographs increasingly seem to be used for live communication instead of for storing moments of life for later recall. What are the implications of this transformation for our quotidian uses of personal photography?

Underlying these three questions is the recurrent issue of control versus lack of control. Part of the digital camera's popularity can be explained by an increased command over the outcome of personal pictures now that electronic processes allow for greater manipulability. Photography in the age of digitization may indeed yield more control over someone's own pictorial identity and personal memory, and yet the flipside is that pictures can also be easily manipulated by everyone who has rudimentary software. A similar paradox can be noticed with regard to the distribution of personal pictures. Although the Internet allows for quick and easy sharing of private snapshots, that same tool also renders them vulnerable to unauthorized distribution. Ironically, the picture taken by the roommate as a token of instant and ephemeral communication may live an extended life on the Internet, turning up in unexpected contexts many years from now. Personal memory insidiously coils with collective memory once pictures are unleashed onto the World Wide Web. As I argue in the last section, the increased malleability of photographic images may suit our need for continuous self-remodeling, but that same flexibility may also lessen our grip on our images' future repurposing and reframing, forcing us to redefine fundamental notions of memory.

Picture Your Self: Remodeling Life

From the early days of photography, its significance as a personal tool derived from its function as a mnemonic aid: without pictures of ourselves, we would most likely lose a sharp idea of what we physically looked like at a younger age. Some theorists claim that personal pictures equal identities (our pictures are us), but this claim appears to understate the intricate cognitive, mental, social, and cultural processes at work in memory and identity formation.[4] Pictures of family and friends are visible reminders of historical appearances, inviting us, as Roland Barthes assumes, to reflect on "what has been."[5] By the same token, personal pictures dictate our autobiographical memory: they tell us, time and again, how we should remember ourselves as younger persons. We remodel our self-image to fit the pictures taken at previous moments in time. Memories are made as much as they are recalled from photographs; our recollections never remain the same, even if the photograph appears to represent a fixed image of the past. And yet, we use these pictures not to fix memory but to constantly reassess our past lives and reflect on what has been as well as what is and what will be. As extensively described in Chapter 2, recollecting is not simply a revisionist project; anticipations of future selves inform retrograde projections, and these mental image maps, in turn, feed a desire to impact external (camera) visions of ourselves.[6]

The role photographs play in the complex construction of one's personal memory and identity has been theorized in cognitive theory as well as in cultural theory, particularly semiotics. Cognitive psychologists have investigated the intriguing question of how photographs can influence personal memories.[7] The human mind actively produces visual autobiographical evidence through photographs, but it also modifies this evidence through pictures—cutting off estranged spouses or throwing away depressing images depicting them when they were still seriously overweight. Research has shown that people are easily seduced into creating false memories of their pasts on the basis of unaltered as well as doctored pictures. In the early 1990s, researchers from America and New Zealand persuaded experimental subjects into believing false narratives about their childhoods, written or told by family members and substantiated by "true" photographs.[8] Over the next decade, these findings were corroborated by experiments in which doctored pictures were used; more than 50 percent of all subjects constructed false memories out of old personal photographs that were carefully retouched to depict a scene that had never happened in that person's life.[9] There is a continuing debate whether it is narratives or photographs (or a combination of both) that trigger most false memories, but the conclusion that people's autobiographical memories are prone to either self-induced intervention or secret manipulation is well established.[10] Not surprisingly, these scientific insights are gratefully deployed in marketing and advertising departments to advance sales by manipulating customers' memories about their pasts and thus influence their future (buying) behavior. What customers recall about prior product or shopping experiences often differ from their actual experiences if marketers refer to those past experiences in positive ways.[11]

The close interweaving of memory, imagination, and desire in creating a picture of one's past has also been subject to theoretical probing by cultural theorists, most notably Roland Barthes. When exploring the intricacies of the camera lucida, Barthes testifies to this complex loop of images/pictures informing desire/memory when describing the discomfort he feels the moment he succumbs to being the camera's object. Having one's photograph taken, as Barthes observes, is a closed field of forces where four image repertoires intersect: "the one that I think I am" (the mental self-image); "the one I want others to think I am" (the idealized self-image); "the one the photographer thinks I am" (the photographed self-image); and "the one the photographer makes use of when exhibiting his art" (the public self-image or imago).[12] Whereas the first two levels represent the stages of mental, internal image processing, the third and fourth levels refer to the external process of picture taking and presentation—the photographer's frame of reference and cultural perspective. In contrast to psychologists, Barthes's semiotic perspective emphasizes that cognition does not necessarily reside inside our brains but extends into the social and cultural realm.

Barthes's exploration of analog photography elucidates how the four image repertoires of self intersect and yet never match. They collide at various moments: at the instant of capturing, when evaluating the outcome or photographed object, or while reminiscing at a later point in time, reviewing the picture. When a picture is taken, we want those photographs to match our idealized self-image—flattering, without pimples, happy, attractive—so we attempt to influence the process by posing, smiling, or giving instructions to the photographer. At a later stage, we can try to change the undesired outcome by selecting, refusing, or destroying the actual print. A photographed person exerts only limited control over the resulting picture. The photographer's choice of frame and angle defines the portraiture, and the referent can be further modified at the stage of development by applying retouching techniques. Barthes obviously feels powerless in the face of the photographer's decisions, lacking control over the picture's referent, which he wants to equal his idealized self. Its fate is in the hands of the photographer who is taking the picture and of the chemical, mechanical, and publishing forces involved in its ultimate materialization. Barthes's discomfort signals a fundamental resentment about his inability to fashion pictures in his own image. Because the four levels never coincide, portrait photographs are profoundly alienating, even to the extent of giving the French philosopher a sense of imposture.

Paradoxically, Barthes perceives a lack of control over his photographed image and imago and yet he feels confident he can exert power over the mental and idealized images entering his mind. According to Barthes, our mental capacities determine which images are allowed to enter our minds and memories. Photos that "work" are those you still remember when you no longer see them. To test the affect of a picture, Barthes suggests that you close your eyes and wait to see if the photographed image is inculcated in the mind—that is, if it conjures up mental images. Barthes's now famous concept of the photograph's *punctum* uses memory as a litmus test for admitting pictures to the mental reservoir. In a similar vein, the autobiographical self selects those photographed images that work the mind, though they do not necessarily "work" to accommodate or feed its idealized self-images. On the contrary, old photographs of self may counteract such projections. For instance, when looking at a group portrait of myself amidst a class of sixth graders, I may identify myself as the one who looks dumb or innocent, or I may point at the stupid dress my mother made me wear to school. In the process of recalling, photographs may block or parcel out specific mental images. We can apparently resist photographs the way we resist memories, not just by literally destroying them but also by simply barring them from entering the mind's eye. The photograph is never a simple conduit but is made to signify something by the mind's ability to frame a picture each time it is (re)viewed.

Barthes's perceived powerlessness over the photographer's perspective and the black box of the camera in relation to the assumed autonomy over his mental images and memories appears entirely plausible, and yet neither perception can go undisputed. The photographed image—the manipulation of one's public imago—has never been outside the subject's influence. Since the late 1840s, commercial portrait photographers have succumbed to their patrons' desire for idealized self-images the way painters did before the advent of photography: by adopting flattering perspectives and applying chemical magic. However, the subject's power over images entering the mind may not be as manageable as it appears. Cultural ideals of physical appearance, displayed through photographs and evolving over time, often unconsciously influence the mind's (idealized) images of self.[13] Control over photographic images is hence not inscribed in the machine's ontology, and neither does the mind have full sovereignty over the images it allows to enter memory. Instead, control over one's self-portrait is a subtle choreography of the four image repertoires, a balancing act in which photographic images enculturate personal memory and a subject's memory evolves through cultural engagement.[14] Individual cognitive editing and processes of image manipulation are intertwined at every level of cultural memory.

Now when we replace the analog camera with a digital one, and laminated photos with pixeled shots, how does this affect the intertwining of mental-cognitive and cultural-material image processes in photography? The attempt to answer this question reveals a conspicuous absence of interdisciplinary research in this area. None of the cognitive studies discussed above pay attention to ways in which individuals use digital photography to manipulate their own personal pictures and memory; the cultural, material, and technological aspects of memory morphing appear strikingly irrelevant to cognitive science. This is striking because scientists often mention how their academic interest in manipulated pictures gains relevance in the face of a growing ubiquitous use of digital photography and its endless potential to reconstruct and retouch one's childhood memories; skills once monopolized by Hollywood studios and advertising agencies are now within the reach of every individual who owns a "digital camera, image editing software, a computer, and the capacity to follow instructions."[15] Indeed, without digital photo enhancement equipment, cognitive psychologists would have a hard time conducting their research on manipulated autobiographical memory in the first place; only with the help of computer paintbrush programs can they make doctored photographs look immaculate. Mutatis mutandis, when turning to cultural theorists for enlightenment, their disregard of psychological and cognitive studies in this area is rather remarkable; semioticians and constructivists typically analyze the intricacies of technological devices to connect them to social and cultural agency.[16] Yet without acknowledging the profound interlacing of mental, technical, and cultural levels involved in digital photography, we may never understand the intricate connection between memory, identity formation, and photography.

It may be instructive to spell out a few significant differences between analog and digital photography in terms of their (cognitive and technical) mechanisms. At first sight, digital photography provides more access to the imaging process between the stages of taking the picture and looking at its printed result. Only seconds after its taking, the picture may be previewed via the camera's small screen. The display shows a tentative result, an image that can be deleted or stored. Because this sneak preview allows the photographer to instantly share the results with the photographed subject, there is room for negotiation: the subject's evaluation of his or her self-image may influence the next posture. A second review takes place at the computer, in which images, stored as digital code, are susceptible to editing and manipulation. Besides selecting or erasing pictures, photo-paint software permits endless retouching of images—everything from cropping and color adjustment to brushing out red eyes and pimples. Beyond the superficial level, one can remove entire objects from the picture, such as unwanted decorations, or add desirable features, such as sharper cheekbones or palm trees in the background.

It is important to be straight about one thing: digitization never caused manipulability or artificiality. Although some theorists of visual culture have earmarked manipulability as the feature that makes digital photography stand out from its analog precursors, history bespeaks the contrary.[17] Retouching and manipulation have always been inherent to the dynamics of photography.[18] What is new in digital photography is the increased number of possibilities to review and retouch one's own pictures, first on a small camera screen and later on the screen of a computer. When pictures are taken by a digital camera, the subject may feel empowered to steer its outcome (the photographed or public image) because he or she may have access to stages formerly "black boxed" by cameras, film roles, and chemical labs. Previews and reviews of the pixeled image, combined with easy-to-use photoshop software, undoubtedly seduce pictorial enhancement. But does this increased flexibility cause the processes of *photographic* imaging and *mental* (or cognitive) editing to further entwine in the construction of autobiographical memory? In other words, does image doctoring become an integral element of autobiographical remembering? Answering this question requires us to include culture in our explanation.

Of course, we have already become used to the prevailing use of the "camera pictura" with regard to the creation of *public* images. Since the 1990s, people no longer expect indexical fidelity to an external person when looking at photographic portraits, particularly those in advertising; almost by default, pictures in magazines, billboards, and many other public sources are retouched or enhanced. Digital stock photography uses public images as resources or input to be worked on by anyone who pays for their exploit.[19] Companies like Microsoft and Getty have anticipated the consequence of this evolution by buying up large stocks of public images, licensing their re-creations, and selling them back to the public domain. From the culturally accepted modifiability of public images it is only a small step to considering your own personal pictures to be mere stock in the ongoing remodeling project of life's pictorial heritage. The impact of editing software on the profiling of one's personal identity and remembrance is evident from many photoblogs and personal picture galleries on the Internet. Enhancing color and beautifying faces is no longer the department of beauty magazines: individuals may now purchase photoshop software to brush up their cherished images. A large number of software packages allow users to restore their old, damaged, and faded family pictures; in one and the same breath, they offer to upgrade your self-image. For instance, VisionQuest Images advertises its packages as technical aids to create a "digital masterpiece of your specification"; computer programs enable you to change everything in your personal appearance, from lip size to skin tint.[20] Examples of individuals who use these programs abound on the Internet. Asian American student Chris Lin, for instance, admits in his photoblog that he likes to picture himself with brown hair; he also recolors the faces of his friends' images to see if it enhances their appearance.[21] Nancy Burson, a New York–based artist and pioneer in morphing technology, attracted a lot of media attention with her design of a so-called Human Race Machine, a digital method that effortlessly morphs racial features and skin colors in pictures of peoples' faces.[22] These instances divulge that the acceptability of photographic manipulation of personal photographs can hardly be separated from the normalized use of enhanced idealized images. Digital doctoring of private snapshots is just another stage in the eternal choreography of the (mental and cultural) image repertoires identified by Barthes.

The endless potential of digital photography to manipulate one's self-image seems to render it the favorite tool for identity formation and personal memory construction. Whereas analog photographs were often erroneously viewed as the still input for static images, digital pictures more explicitly serve as visual resources in a lifelong project to reinvent one's self-appearance: they become "living pictures" amenable to infinite change. Software packages supporting the processing of personal photographs often bespeak the digital image's status as a liminal object; pixeled photographs are touted as bricks of memory construction, as software is architecturally designed with future remodeling in mind.[23] As Canadian design scholar Ron Burnett eloquently phrases it: "The shift to the digital has shown that photographs are simply raw material for an endless series of digressions. . . . As images, photographs encourage viewers to move beyond the physical world even as they assert the value of memory, place, and original moments."[24] Digital photography, in other words, advances the concept of autobiographical remembrance as a mixture of memory and desire, of actual pictures and idealized images, of constantly evolving input and output.

I am not saying, though, that with the advent of digital photography people all of a sudden feel more inclined to photoshop their personal pictures stored in the computer. Nor am I arguing that mental imaging processes change as a result of having more access to intermediate layers of photographic imaging. My point is that the condition of modifiability, plasticity, and ongoing remodeling equally informs—or enculturates—all four image repertoires involved in the construction of personal memory. The condition of plasticity and modifiability, far from being exclusive to memory, resounds in diverging cultural, medical, and technological self-remodeling projects. Ultrasound images of fetuses—sneak previews into the womb—stimulate intervention in the biological fabric, turning the fetus into an object to be worked on.[25] Cosmetic surgery configures the human body as a physical resource amenable to extreme makeovers; before-and-after pictures not only structure subjective self-consciousness, but upon entering the public image repertoire, they concurrently normalize intervention in physical appearance. The most remarkable thing about before-and-after pictures abounding on the Internet and on television these days is that they do not promote perpetual modification of our pictures to portray a better self, but rather they advertise the potential to modify our bodies to match our idealized mental images. Contemporary notions of body, mind, appearance, and memory seem to be equally informed by the cultural condition of perpetual modification; our new tools are only in tune with the mental flexibility to refashion self-identity and to morph corporeality.

The question whether changing concepts of personal memory have followed from evolving technologies or the other way around is in fact beside the point. What is more important is to address how the new choreography of image repertoires operates in a social and cultural climate that increasingly values modifiability and flexibility, and whether this climate indeed allows more individual control over one's own image. But to understand this larger picture, we first need to look at the sociotechnical aspects of photography. The camera is not solely used as a tool for identity construction and remembrance; the apparatus of personal photography also produces material keepsakes that embody routines and generate commercial products.

Digital Photographs as Material Keepsakes

Photographs are by far the most cherished memory objects in someone's personal heritage, probably more so than videos, diaries, and tapes combined; they owe their special status to a strange mix of material and immaterial preciousness, a blend of "thingness" and performativity, and a combination of domestic and commoditized value. The power of photographic images stems from their being material realities in their own right, while they are coveted as symbolic personal possessions. People may store their pictures as ongoing collections in shoeboxlike containers, or they may order them carefully in albums to present them as finished narratives of a specific period in their lives. Different storage frames engender different performative routines of showing pictures: a box with slides implies a different presentational setting than a photo album. And although personal photographs are commonly considered domestic items without any monetary value, they also form the core product of an extensive commercial branch.[26] From the development of film rolls to the creation of digital picture management systems, that mythical space known as private life, captured in personal photographs, is pervaded by industrial services and products, not only ensuring continuous profits but also guaranteeing a firm grasp on their ideological framing. The photographic image, as Don Slater argues, takes its shape and force from the "mélange of domesticity, consumerism and leisure."[27]

When considering current trends spurred by digitization, we may witness an interesting metamorphosis in the photographic object's materialization as well as in its performative nature and its commoditized essence. Let me subsequently illustrate each of these transformations. Considering a photograph's materiality, it may be instructive to once again compare digital and analog photography, now in terms of picture processing. In film cameras, the output commonly refers to two things—negative images on celluloid strips and laminated paper prints—the status of which differs considerably. Negatives are not kept to be looked at but to ensure future reproduction; like slides, their value as objects lies in their reproducibility and "projectionability." Although the negative of a photograph can guarantee infinite reproduction, its potential to be retouched is rather limited and the process cumbersome. Laminated prints are not only images, but they are the finished products of chemical processes, paper objects to be preserved and presented.[28] Shoeboxes or photo albums hold them together as collections, and the tangibility of these paper products constitutes an important part of their value as pictorial heritage.

The new materiality of digital photographs is often referred to as virtual or intangible, but such denomination erroneously connotes immateriality. In digital photography, celluloid strips are replaced by invisible binary codes, stored on computer hard disks or on external storage mediums. Coded images hold a material value similar to their celluloid counterparts, even if they cannot be looked at unless they are turned into pixeled objects. Once they appear on the screen, they are visible entities, even if they have a different feel to them than slides or laminated prints. It is precisely this pixeled quality or the bright backlit screen that may cause its specific affect. In addition to being reproducible and projectionable, virtual images are also malleable and versatile: they can be endlessly and effortlessly retouched without losing quality, and they can be printed on paper as well as on mugs or T-shirts. The versatility of coded pictures supports the managing act of personal identity construction: we may now store infinite numbers of single pictures on our digital media and re-present them in multiple different formats, such as photoblogs, Powerpoint presentations, or multimedia configurations, without compromising their quality.

Along with their materiality, the performative nature of pictures changes upon moving to a digital platform; so far, little attention has been given to these objects' agility or their illocutionary force.[29] Digital presentational formats of pictures engender different performative uses than analog ones. Instead of framed pictures showcasing loved ones on someone's office desk, we can now expect these images to light up on desktop screensavers. Exhibiting the ritual highlights of family life in a professional environment, the illocutionary force of the screensaver, though, diverges from the framed picture's message: rather than a mere ornament on permanent display, the reminder of home is effectively switched on when taking a break from the computer. Virtual objects do not automatically replace printed pictures; on the contrary, they are often exhibited complementary to photographic prints, or they are used to personalize someone's computer, desk, or workspace.[30] Photos sent as e-mail attachments to show a toddler's latest achievements may have been distributed with the same intent as laminated pictures enclosed with a letter, but that does not mean they convey the same message. The connotation of e-mailing pictures as attachments is one of transience rather than of permanence, a mere update rather than a record. Along similar lines, showing a slide show on a laptop is a performative act ardently reminiscent of an old-fashioned slide show, and yet the casualness of the gesture implied in the laptop presentation sharply contrasts the cerebral efforts involved in setting up the slide projector, pulling out the screen, and sorting through the slide racks.

Digital photography elicits a performativity and materiality that deserves to be evaluated in its own right, not just as a virtualization of former analog products and practices. The virtual image, as a screen object, is gradually acquiring a new status beside the laminated object, which in turn adjusts its meaning to suit this pixeled neophyte. We can see this also in relation to methods of storing and presenting personal picture collections. Software engineers increasingly begin to realize that the design of picture management systems requires a profound understanding of why and how users interact with their pictures: the acts of storing pictures in a shoebox or sticking them into albums cannot simply be transposed onto digital platforms.[31] In terms of hardware, the single-purpose camera for taking still pictures gives way to multifunctional appliances, combining the camera function with the personal digital assistant (PDA), the mobile phone, MP3 players, and global positioning devices. But the so-called camera phone also permits entirely new performative rituals, such as shooting a picture at a live concert and instantly e-mailing the image to a friend. Emerging digital tools are thus deeply affecting the way people socialize and interact and, by extension, the way they maintain relationships and consolidate them into personal memory.

Because personal photography is the pivot of a large industrial branch, commercial stakes in the digital refurbishing of material keepsakes are high. The logic of consumption and replenishment very much underlies digital photography, hence the huge economic interest in creating and tracing new material and performative uses of photography. Hardware industries are gradually shaking off their chemical and laboratory divisions, instead concentrating on corporate strategies that concurrently target digital and paper image processing. Software developers respond to the camera phone trend by integrating photography into multifunctional scripts for online communication and shareware. Cameras and related paraphernalia still support the demand for printed snapshots—partly by bringing the printing process into the do-it-yourself market—while also inscribing the new preference for sharing pictured experience into their technological scripts. So-called dock-and-share systems cater to people's dual desire for instant gratification and instant notification.

In the analog days, the wish to have picture viewing virtually coincide with real-time experience could only be achieved by using Polaroid or Instamatic cameras that ejected instant laminated pictures. The drawbacks of Polaroids—they were very expensive and produced rapidly fading print images without negatives—made them impractical as memory tools. Hence, their success was mostly limited to (commercial) exploits where pictures were sold as signals of presence: photographers offering instant snapshot services at entertainment venues such as roller coasters and canal cruises or distributing thank-you pictures to wedding guests as they leave the party. Digital dock-and-print systems bring some of these formerly commercial applications within the reach of individual consumers; without the drawbacks attached to Instamatics and Polaroids, the rituals of pictorial thank-you notes and personalized invitations may rise to a new normative level. Besides banking on instant gratification, digital camera and docking systems increasingly cater to the desire for instant notification. Whereas Kodak's Brownie, in 1901, popularized the camera with its slogan "You push the button, we do the rest," the company's newest camera, Easy Share, highlights its most prominent feature: a button to automatically e-mail a picture upon docking the camera. Instant sharing and self-service, rather than easy use and full service ("we do the rest"), appear the new selling points of digital pocket cameras.

The logic of consumption still underpins the need for more technology and its industrial spin-offs, although we can recently detect a shift from easy-to-use cameras and ready-made laminated objects to camera paraphernalia and software supporting instant materialization and exchange; however, expectations that digital cameras will result in more pictures but fewer actual prints may well turn out to be another paperless-office myth. This technological and material transformation seems to come with an inherent bias toward the communicable and the disposable, at the expense of permanence and durability. Whereas laminated pictures are meant to be stored as keepsakes, coded or screen images tend to be assigned a temporary (exchange) value, always amenable to recycling or reframing. The laminated print is an object to hold on to, whereas the digital picture appears to be an object to work on and distribute.

However, conclusions on the changing material and technological nature of personal photography are inevitably bound up with questions concerning photography's transforming sociocultural use. The digital camera and its resulting snapshots give rise to newly accentuated uses and ritualized practices. Easy manipulation and sharing are not simply new technological features; they reflect and construct a desire to control and direct identity formation. And yet, if digital photography indeed bodes a preference for networked distribution and manipulation—consolidating its penchant toward the transient and ephemeral—how does this affect its use as a tool for remembrance?

Digital Photography as Communication and Experience

When personal photography came of age in the nineteenth and twentieth centuries, it gradually emerged as a social practice that revolved around families wanting to save their memories of past experiences in material pictorial forms for future reference or communal reminiscing. Yet even in the early days of photography, social uses complementary to its primary function were already evident. Photography always also served as an act of communication and as a means to share experience. As Susan Sontag argued in 1973, the tourist's compulsion to take snapshots of foreign places reveals how taking pictures can become paramount to experiencing an event; at the same time, communicating experiences with the help of photographs is an integral part of tourist photography.[32] Notwithstanding the dominance of photography as a family tool for remembrance and reminisce, the communicative function was immanent to photography from the moment it became popular as a domestic technology. In recent years, there have been profound shifts in the balance between these various social uses: from family to individual use, from memory tools to communication devices, and from sharing (memory) objects to sharing experiences. I subsequently elucidate each of these profound shifts.

The social significance and cultural impact of personal photography grew exponentially in the past century: by the early 1970s, almost every

American and western European household owned a photo camera. Long before sociologists and anthropologists began to acknowledge the significance of photography as a cultural rite of family life, Sontag took on the ethnographer's cloak and described its meaning as a tool for recording family life: "Through photographs, each family constructs a portrait chronicle of itself—a portable kit of images that bears witness to its connectedness."[33] By taking and organizing pictures, individuals articulate their connections to, and initiation into, clans and groups, emphasizing ritualized moments of aging and of coming of age. Cameras go with family life, Sontag observed: households with children are twice as likely to have at least one camera as households in which there are no children. Photography did not simply reflect but constituted family life and structured an individual's notion of belonging. Quite a number of sociological and anthropological studies have scrutinized the relationship between picture taking, organizing, and presenting photographs on the one hand and the construction of family, heritage, and kinship on the other.[34]

Over the past two decades, the individual has become the nucleus of pictorial life. In her ethnographic study of how people connect personal photographs to memory and narration, anthropologist Barbara Harrison observes that self-presentation—rather than family re-presentation—has become a major function of photographs.[35] Harrison's field study acknowledges a significant shift from personal photography as a tool bound up with memory and commemoration toward pictures as a form of identity formation; cameras are used less for the remembrance of family life and more for the affirmation of personhood and personal bonds.[36] Since the 1990s, and most distinctively since the beginning of the new millennium, cameras increasingly serve as tools for mediating quotidian experiences other than rituals or ceremonial moments. Partly a technological evolution pushed by market forces, the social and cultural stakes in this transformation cannot be underestimated. When looking at current generations of users, researchers observe a watershed between adult users, large numbers of whom are now switching from analog to digital cameras, and teenagers and young adults, who are growing up with a number of new multifunctional communication and media devices.[37] The older group generally adheres to the primacy of photography as a memory tool, particularly in the construction of family life, whereas teenagers and young adults use camera-like tools for conversation and peer-group building.

This distinctive swing in photography's use also shows up in ethnographic observations of teenagers' patterns of taking and managing pictures. One American study focusing on a group of teens between fourteen and nineteen years of age reports a remarkable incongruence between what teenagers say they value in photography and how they behave: most of them describe photos as permanent records of their lives, but their behavior reveals a preference for photography as social communication.[38] Showing pictures as part of conversation or reviewing pictures to confirm social bonds between friends appears more important than organizing photos in albums and looking at them—an activity they consider their parents' domain. Photos are shared less in the context of family and home and more in peer-group environments: schools, clubs, friends' houses. The study notes how teens regard pictures as circulating messages, an interactive exchange in which personal photographs casually mix public images, such as magazine pictures, drawings, and text.[39] In the past three years, photoblogs have become popular as an Internet-based technology—a type of blog that adds photographs to text and hyperlinks in the telling of stories. Photoblogs have an entirely different function than photo albums, but they are also different from the lifelogs described in Chapter 3. Photobloggers want to promote photographs as more valuable than words, and they profile themselves in their pictures.[40]

Whereas their parents invested considerable time and effort in building up material collections of pictures for future reference, youngsters appear to take less interest in sharing photographs as *objects* than in sharing them as *experiences*.[41] The rapidly increasing popularity in the use of camera phones supports and propels this new communicative deployment of personal photography. Pictures distributed by a camera phone are used to convey a brief message, or merely to show affect. Connecting and getting in touch, rather than reality capturing and memory preservation, are the social meanings transferred onto this type of photography. Whereas parents and/or children used to sit on the couch together flipping through photo albums, most teenagers today consider their pictures to be temporary reminders rather than permanent keepsakes. Phone photography gives rise to a cultural form reminiscent of the old-fashioned postcard: snapshots with a few words attached that are mostly valued as ritual signs of (re)connection.[42] Like postcards, camera phone pictures are meant to be discarded after they are received.

Not coincidentally, the camera phone merges oral and visual modalities—the latter seemingly adapting to the former. Pictures become more like spoken language as photographs are turning into the new currency for social interaction. Pixeled images, like spoken words, circulate between individuals and groups to establish and reconfirm bonds. Sometimes pictures are accompanied by captions that form the missing voice to explain the pictures. For instance, a concert visitor takes a camera phone picture of her favorite band, adds the caption "awesome," and immediately sends off the message to her friends back home. Camera phone pictures are a way of touching base: Picture this, here! Picture me, now! What makes camera phones different from the single-purpose camera is the medium's verbosity—the inflation of images inscribed in the apparatus's script. When pictures become a visual language channeled by a communication medium, the value of individual pictures decreases, whereas the general significance of visual communication augments. A thousand pictures sent over the phone may now be worth a single word: see! Taking, sending, and receiving photographs are real-time experiences, and like spoken words, image exchanges are not meant to be archived.[43] In their bounty, photographs gain value as "momentos" while losing value as mementos.

Even though photography may still capitalize on its primary function as a memory tool for documenting a person's past, we are witnessing a significant shift, especially among the younger generation, toward using it as an instrument for interaction and peer bonding. Digitization is not the cause of this trend; instead, the tendency to fuse photography with daily experience and communication is part of a broader cultural transformation that involves individualization and intensification of experience. The emphasis on individualism and personhood at the expense of family is a social pattern whose roots can be traced back as far as the late 1960s and early 1970s. The intensification of experience as a turn-of-the-millennium economic and social force has been theorized most acutely by American economists Joseph Pine and James Gilmore; commercial products are increasingly marketed as memorable experiences engaging all five senses—sight, sound, touch, taste, smell—and packaged with snappy themes, so as to prolong the contact zone between product and consumers.[44] Digital photography is part of this larger transformation in which the self becomes the center of a virtual universe made up of informational and spatial flows; individuals articulate their identity as social beings not only by

taking and storing photographs to document their lives but also by participating in communal photographic exchanges that mark their identity as
interactive producers and consumers of culture.

From the above observations, it is tempting to draw the conclusion
that digital cameras are moving away from their prime functions as memory tools, instead becoming tools for identity formation, communication,
and experience. If photographs were always a medium for remembering
scenes and objects from the past, digital cameras particularly encourage
users to imagine and invent the present. Digital personal photography gives
rise to new social practices in which pictures are considered visual resources
in the microcultures of everyday life.[45] In these microcultures, memory
does not so much disappear from the spectrum of social use as it takes on a
different meaning. In the networked reality of people's everyday life, the
default mode of personal photography becomes sharing. However, few
people realize shared experience almost by definition implies distributed
storage: personal live pictures distributed through the Internet may remain
there for an indefinite period, turning up in unforeseen contexts, reframed
and repurposed. A well-known example may clarify the meaning of distributed memory and demonstrate the intertwined meanings of personal and
collective cultural memory: the Abu Ghraib pictures.

In May 2004, a series of the most horrific, graphic scenes of torture
and violence used by American guards stationed at the Abu Ghraib prison
against Iraqi detainees appeared in the press, and they were subsequently
disseminated through the Internet.[46] Most pictures were made by prison
guards and frequently featured two lower ranked members of the armed
forces, Charles Graner and Lynndie England; they often posed giving the
thumbs-up in front of individual or piled up prisoners who invariably
showed signs of torture or sexual assault. The hundreds of pictures taken
by prison guards of detainees communicate an arduous casualness in the
act of photographing. Clearly, these picture were made by digital cameras
(or camera phones) deployed by Army personnel as part of their daily
work routines—perfectly in tune with the popular function of photography as a ritual of everyday communication. As Sontag poignantly describes
in her essay on the case: "The pictures taken by American soldiers in Abu
Ghraib reflect a recent shift in the use made of pictures—less objects to be
saved than messages to be disseminated, circulated. A digital camera is a
common possession among soldiers. Where once photographing war was

the province of photojournalists, now the soldiers themselves are all photographers—recording their war, their fun, their observations of what they find picturesque, their atrocities—and swapping images among themselves and e-mailing them around the globe."[47] Intentionally taken to be sent back home as triumphant trophies or to be mailed to colleagues, the pictures were a social gesture of bonding and poaching. Some pictures allegedly served as screensavers on prison guards' desktops, a sign of their function as office jokes to be understood by insiders only. The casualness and look-at-me-here enunciation of the Abu Ghraib photographs, conveyed by the uniformed men and women whose posture betrayed pride as if they had just caught a big fish, connotes the function of these pictures as symbolic resources for communication. Their makers never meant for these pictures to be objects of lasting memory.

And yet, this is exactly how they ended up in the collective memory of the American people and the world. Once interceded and published in newspapers and on television worldwide, they were reframed as evidence of the Army's abhorrent behavior as torturers posing triumphantly over their helpless captives. The Abu Ghraib pictures became evidence in a military trial that incriminated the perpetrators responsible for the abuse shown in the pictures but acquitted the invisible chain of command that obviously condoned such behavior. Perhaps most telling was the military's response to the Abu Ghraib debacle. Rather than condemning the practice depicted by the images taken, the military subsequently ordered to ban personal photography from the work floor; pictures made for private use may no longer be taken outside penitentiaries. The incident resulted in stricter communication regulations as well as a prohibition against taking and distributing personal photographs on military premises.

Ironically, pictures that were casually mailed off as ephemeral postcards, meant to be thrown away after reading the message, became a permanent engraving in the consciousness of a generation; pictures sent with a communicative intent ended up in America's collective cultural memory as painful visual evidence of its military's hubris. The awareness that any picture unleashed on the Internet can be endlessly recycled may lead to a new attitude in taking pictures: anticipating future reuse, photographs are no longer innocent personal keepsakes, but they are potential liabilities in someone's personal life or professional career. The lesson learned from the Abu Ghraib pictures—beyond their horrendous political message—is that personal digital photography can hardly be confined to private grounds; embedded in networked systems, cultural memory is forever distributed, perpetually stored in the endless maze of virtual life.

Digital Photography and Image Control

The digital evolution that has shaped personal photography is anything but an exclusive technological transformation. "Picturing the self" appears to be an *embodied* act, *enabled* by the newest digital tools and *embedded* in the new cultural practices these tools bring along. The shift in use and function of the camera as a tool for identity formation and communication seems to suit a more general cultural condition that may be characterized by terms such as "manipulability," "experience," "versatility," and "distributedness." This cultural condition has definitely affected the nature and status of photographs as building blocks for personal identity and as material input for acts of communication. Even if the function of memory capture persists in current uses of personal photography, its reallocated significance reverberates crucial changes in our contemporary cultural condition. Returning to the issue of power, it is difficult to conclude whether digital photography has led to more or less control over our personal images, pictures, and memories. The choreography of image repertoires, blending mental and cultural imaging processes, not only seems to reset our control over pictures and memory but also implies a profound redefinition of the very terms.

Photographs, as mediated memories, could never be qualified as truthful anchors of personal memory; yet since the emergence of digital photography, pictorial manipulation seems to be a default mode rather than an option. To some extent, the camera allows more control over our memories, handing us the tools to brush up and reinvigorate remembrances of things past. Presently, photography allows subjects take some measure of control over their photographed appearance, inviting them to tweak and reshape the referent. As stated earlier, digital photography is not the cause of memory's transformation, nor is personal memory directly the result of technical changes. The digital camera derives its revamped application as a memory tool from a culture where manipulability and morphing are commonly accepted conditions for shaping personhood. Flexibility and morphing do not apply exclusively to pictures as shaping tools for personal memory but also apply to bodies. Memory, like photographs and bodies, can now be made picture perfect; memory and photography change in conjunction, adapting to contemporary expectations and prevailing norms. Our photographs tell us who we want to be and how we want to be remembered. Just as before-and-after photos on the Internet or on television normalize the need for cosmetic surgery, the abundance of editing tools available will make the inclination to brush up our past selves even more acceptable. Personal photography may become a lifelong exercise in revising past desires and adjusting them to new expectations. Even if still a memory tool, the digital camera is now pushed as an instrument for identity construction, allowing more shaping power over autobiographical memories.

And yet, this same manipulative potential that empowers people to shape their identity and memory may be also used by others to reshape that image. The consequence of digital technology is that personal pictures can be retouched without leaving traces and be manipulated regardless of ownership or intent of the original picture, as evidenced at the beginning of this chapter by the anecdote about the student who was unpleasantly surprised to find a doctored picture of her and her friends electronically distributed to (anonymous) recipients. Personal photographs are increasingly pulled out of the shoebox to be used as public signifiers. Pictures once bound to remain in personal archives increasingly enter the public domain, where they are invariably brushed up or retouched to (retro)fit contemporary narratives. It is quite plausible to see personal pictures emerge in entirely different public contexts, either as testimony to a criminal on the run, as a memorial to a soldier who died in the war, or as evidence of a politician's excessive alcohol use in college.[48] Like the subjects who were shown altered pictures in the psychologists' experiments, we may be unable to determine whether they are true or false: is it memory that manipulates pictures, or do we use pictures to create or adjust memory? The digital age is setting new standards for remembrance and recall: the value of the terms "true" and "doctored" in the context of memory will have to be reconsidered accordingly.

The ability of photographic objects to evoke personal memories, is increasingly giving way to its communicative and experiential uses. In addition to photographs' function as material keepsakes, once primarily intended for veneration or ritual use—stored in the family archive or exposed on the walls of the home gallery—photographs metamorphose into virtual objects of exchange and versatile coded artifacts. Once treated as memory objects imbued with private connotations, the casual dissemination of digital pictures signifies a new ephemeral meaning. Pictures taken by a camera phone, meant as expendable enunciations to be shared with co-workers, have a distinctly different discursive power than the framed black-and-white ancestor photographs on the wall. We may now take pictures and electronically distribute them to a number of known and anonymous recipients, enabling instant notification and gratification. Networked systems define new presentational contexts of personal pictures, as sharing pictures becomes the default mode of this cultural practice. In many ways, digital tools and connective systems expand control over an individual's image exposure, granting more power to present and shape oneself in public.

However, the flipside of this increased versatility is actually a loss of control over a picture's framed meaning: pictures that are amenable to effortless distribution over the Internet are equally prone to unintended repurposing. But because the framing of a picture is never fixed for once and for all, each rematerialization comes attached with its own illocutionary meaning and each reframing may render the original purpose unrecognizable. So even if taken with a communicative use in mind, a picture may end up as a persistent object of (collective) cultural memory—as evidenced by the Abu Ghraib pictures. The consequences of reframing and repurposing are particularly poignant when pictures move seamlessly between private and public contexts. Of course, this risk is never the direct implication of photography's digital condition, but it cannot be denied that digitization has made reframing in electronic media much easier and smoother. Distributing personal pictures over the Internet or by camera phone, which is now a common way to communicate, intrinsically renders private pictures into public property and therefore diminishes one's power over their presentational context.

Image control is still a pressing concern in the debates over personal photography in the digital age, even if the parameters for this concern have substantially shifted, adapting to new cognitive mindsets as well as to new technological, social, and cultural conditions. We may hail the increased manipulability of our self-image due to digital photography while at the same time we resent the loss of power over our pictorial framing in public contexts.[49] The enhanced versatility and multipurpose uses of digital pictures facilitate promotion of one's public image and yet also diminish control over what happens once the pictures become part of a networked environment, which changes their performative function each time they are retrieved. Due to this networked condition, the definition of personal cultural memory is gravitating toward a distributed presence. We can no longer keep the lid on the shoebox we used to store in our attic: its pictorial contents will increasingly spill out into the virtual corners of the World Wide Web, where it seamlessly blends in with our collective pictorial heritage. Once again, pictures of life will become living pictures—even if unwittingly.

6

Projecting the Family's Future Past

Upon his return from work, a colleague of mine was buoyantly greeted by his ten-year-old daughter. She begged him to fetch his camcorder and come to her room, where she was playing with her sister—they were performing a karaoke of sorts in which they combined song and dance with typical kids' spells of laughter and fun. "You need to tape us because when we become famous they may show this on television," his daughter explained with a sense of urgency. The children's motivation for being filmed betrays a sophisticated reflexivity of the camcorder as a tool for producing future memories. Even at a young age, children keenly apprehend the pliability of mediated experience; their father's film is not simply a registration of present fun activities, but it is also a conscious steering of their future past. The camcorder constructs family life at the same time and by the same means as it constructs our memory of it; whereas the camcorder registers images of private lives, in the context of television these images may help shape (future) public identity. The children's awareness was most likely triggered by contemporary television programs—anything from so-called reality TV and lost-relative quests to dating shows and celebrity interviews—that deploy home video footage to represent a person's past life.

This scene helps articulate a few intriguing questions on home movies as a means to study the connection between individuals and family, media and memory. Can filming the family be considered an act of cognition—minds instructing instruments to manufacture desirable images of how one wants to remember his or her family in the future?[1] Are home movies registrations of actual or real families, enabled by the various generations of technological equipment? And what if home movies become cinematic constructions of past family life—footage shot and later reassembled into a common cultural format? Filmed families constitute fascinating windows into cultural memory, illustrating how the horizontal axis of relational identity and the vertical axis of time intersect in memory acts and products. As the children in the above scene perfectly understand, video footage may serve to actually steer one's history as it is actively deployed to shape a family's future past. Home movies remade into cultural products like fiction films, documentaries, websites or television programs can hardly be considered mere pictorial representations of family at a certain moment in time; as mediated memories, they are audiovisual constructions of hindsight. The goal of this chapter is to examine family movies as co-productions of mind, technology, and culture: to what extent are the concepts of "home movie" and "family films" mental projections enabled by media technologies and embedded in sociocultural practices?

First, embodied memory as it relates to home movies is examined. Neurobiologists and philosophers of mind have theorized the intimate relationship between the mind's eye and the camera's eye from various angles. Neurobiologists use movies, screens, and cameras as metaphors to describe the intricate mechanism of human consciousness, whereas some philosophers take them to be more than metaphors. Gilles Deleuze, for instance, suggests the inseparability of the human brain and the movie screen. Films are imagined in the brain, a process that involves a convergence of mental projections and technological scripts; projecting the future and capturing the present are closely intertwined activities of memory. Following the footsteps of Deleuze, Mark Hansen upgrades Deleuze's notion of filmed memories to the digital age, arguing how this convergence takes place at the intimate junction of body and technology.

Beyond the "embodied" perspective, a social constructivist angle is introduced, directing the inquiry into mediated memories from the intersection of technology and culture. As James Moran argues, home movie technologies can hardly be separated from social contexts; notions of family change in conjunction with the technical tools we use to capture families in the private sphere of home. The movie camera, the video camera, and more recently the digital camcorder have been used to capture the routines of everyday life. At the same time, "real families" transpire in cultural products everywhere. Families played out in television series shape our mental concepts of family interaction, and home movies provide input for media products. In the age of camcorders, webcams, and multimedia productions, family lives are becoming an insidious part of our combined mental, technological, *and* cultural fabric of memory. We need to differentiate between these three levels in order to understand how, in the digital age, filmed family life always involves remembrance, fabrication and projection. Various examples—a feature film, a documentary, and websites—will illustrate this combined theoretical approach. From the analyses of contemporary mediated memories, I argue that the future of memory will be determined by our tools for reconstruction as much as by our imaginative capacities.

Future Memories as Projections of Family

Let us return for a moment to the scene in which the children demanded their playful activities to be taped by their father's camera. The youngsters were keenly aware of how video footage potentially steers their public image as they grasp the essence of raw images serving as input for memories that have yet to be shaped. But could such awareness be the result of their minds' projections or, perhaps more likely, is it the result of the movie camera's infiltration into their mindsets—the result of the ubiquitous presence of camcorders in their everyday lives? Mind and technology, as I argued in earlier chapters, are closely interwoven in our projections and memories of self. The desire to identify oneself as belonging to a family may be deeply implicated in the private camera as a mnemonic tool—to save visual evidence of family life for later reference—but it may just as well be a function of the brain to funnel conscious perceptions of family into desirable or idealized (moving) images. I succinctly explore various hypothetical angles to account for the complex interrelation between mind and audiovisual technology in the construction of personal cultural memory.

Some neurobiologists who study the physiological mechanisms of autobiographical memory concentrate on the brain as an explanatory framework, and choose to ignore the constitutive function of technology or culture in the process of remembering. When neurobiologist Antonio Damasio speaks of memory as a form of consciousness, he defines it as a two-tired problem. First, he wants to know how the brain turns neural patterns into explicit mental patterns called "images" (see Chapter 4). Multisensory qualia, such as the tone of a violin, the blueness of a sky, and the taste of apple pie, are translated by the brain into mental image maps—image narratives Damasio calls "movies-in-the-brain." Second, in producing those mental images, the brain also engenders a sense of self in the act of knowing; all perceptions are the unmistakable mental property of an automatic owner who concurrently constructs images of objects and of self.[2] Obviously, when Damasio speaks of a movie-in-the-brain, he uses the term metaphorically: as if the brain were both a camera, a movie screen, a filmed production, and a moviegoer. Without a proper analogy, it seems impossible to explain the complexity of the brain's involvement in configuring a sense of self over a period of time. However, the use of this metaphor presents a peculiar paradox: apparently, we need a cultural metaphor (movie) to imagine a physical process (memory, consciousness), whereas actual movies are ultimately the result of a complex brain-machine network involved in film production (scripting, directing, camera work, editing, watching, and so forth)—a subject addressed in the next paragraph. Damasio's theory is understandably oblivious of actual movies as input for actual brains; neither does he account for the role of the camera or other media technologies in equipping the mind's construction of images. The pair brain/mind is hierarchically off set from the pairs technology/materiality and cultural practices/forms; the latter two are mere conceptual aids in the neurobiological theory of movies-in-the-brain.

It is rather interesting to compare Damasio's inquiry into memory as a form of consciousness with Deleuze's philosophical reflections on cinema, memory, and time.[3] Building on Henri Bergson's conjectures in *Matter and Memory*, Deleuze has theorized the internalization of the film camera in the human mind to explain memories as filmic projections of the present. The "matter" of memory, according to the French philosopher, emerges at the intersection of brain/mind and technology/materiality. In his book *The Time-Image*, Deleuze highlights the intimate relation between memory and cinema—between moving images in the mind and moving images on the screen. When Deleuze suggests that "the brain *is* the screen," he does not mean this metaphorically but literally: recollection is inherently defined by the input of actual moving images, which are always partly constructions of the brain.[4] Whereas Damasio's term "movie-in-the-brain" implies a figural equation (to understand the brain's mechanism *in terms of* film productions), Deleuze explicitly connects cognitive mechanisms to the movement-image of cinema. Cinema is as much a production of the individual mind as it is a production of a mechanical apparatus. Echoing Bergson, Deleuze distinguishes between different categories of images in motion: the perception-image, the affection-image, and the action-image.[5] "External images" act upon the mind and become "internal images," moving images that stir action or affection, converging into experience.

Memory, in Deleuze's concept, is never a retrieval of past images but always a function of the present, a function that embodies the essential continuation of time. Body and cinema are part of the same organic, connective system: just as the body constantly renews itself molecularly, images never remain the same when processed in an individual's mind. In other words, moving images produced by film are input for the brain, always resulting in updated output. Such a dynamic concept of movement-image sharply contrasts semiotic theories that consider film footage as representations or signs. "Perception-images" of the present determine how actual images of the past are interpreted, and yet, both are inevitably injected with projections of the future: idealized images, virtual images, desire. It is instructive to quote Deleuze's words in full here: "But instead of a constituted memory, as function of the past which reports a story, we witness the birth of memory, as function of the future which retains what happens in order to make it the object to come of the other memory. . . . [M]emory could never evoke and report the past if it had not already been constituted at the moment when past was still present, hence in an aim to come. It is in fact for this reason that it is behavior: it is in the present that we make a memory, in order to make use of it in the future when the present will be past."[6] Films are imagined in the brain—a process involving the convergence of mental projections and technological scripts. If the past is a filmic product of the present, so is the future; according to Deleuze, memory is always in a "state of becoming."

It takes little effort to translate this part of Deleuze's theory into the mediated memories model outlined in figure 2 (see Chapter 2). Applied to the specificity of home movies, one could argue that the film apparatus is inseparable from the individual who deploys the camera to articulate a sense of connection between self and family. As a memory object, a home movie changes meaning each time it is seen. The act of memory also includes the actual shooting of the movie; the later use of home movies and video footage—even if unspecified—is already anticipated at the moment of shooting. In addition, home movies are never simply found footage of the past: each time they are reviewed or recycled, they are edited by the brain. Moving images, first shot and later edited, project the intricate time-movement of mental recollection so characteristic of the human mind; the mind's tendency to impact future remembrance is implicated in technologies of memory, particularly the movie camera. Deleuze's philosophical reflections on memory and cinema stress the interdependent connections between the functions of the brain/mind in the act of memory and the technology/materiality of the memory object. Applying his theory to specific (fiction) films, Deleuze minutely analyzes how the brain is always involved in articulating moving images produced by the cinema apparatus, an apparatus that only works because the anticipation of mental images is part of its technological script.

However, as we move into an age when the cinematic apparatus and the video image are gradually replaced by the multimedia apparatus and the virtual image, Deleuzian philosophy needs updating in several respects. As Mark Hansen contends, processes of digitization and virtualization call for a new concept of embodiment—the body's relation to image and its affects. Whereas Deleuze still accepts a distinction between perception and simulation—between external and internal images that stir action or affection, converging into experience—Hansen argues that our bodies, brought into contact with the digital image, experience the virtual through affect and sensation rather than through techniques, forms, or aesthetics.[7] Drawing examples from digital art works and virtual reality environments, Hansen counters Deleuze's neuroaesthetics and "cinema of the brain" with a concept in which "the brain is no longer external to the image and is indeed no longer differentiated from an image at all."[8] Rather than talking about an affect caused by a technological (cinematic) apparatus, Hansen considers the digital apparatus to be an integral part of a new embodied experience. Digital technologies call for an approach to cinematic hindsight that privileges the bodily basis of vision; the mind, according to Hansen, filters the information we receive to create images of the past, instead of simply receiving images as preexisting technical forms.

Both Deleuze and Hansen agree that family portraits captured in moving images are never simply retrospectives—found footage as relics of the past—but they are complex constructions of mental projections and technological substrates. Deleuze in particular pays little attention to the sociocultural forces involved in filmic productions.[9] And yet, it is precisely this triangle of forces that seems to account for the richness of family movies as objects of cultural analysis. The social practice of (home) moviemaking and the impact of conventional cultural forms on our understanding of moving images appear to play an equally constitutive role in the construction of cinematic hindsight. A concrete example may help illustrate both the relevance and shortcomings of this techno-embodied perspective, instead arguing for the comprehension of movies as collaborations of mind, machine, and culture.

It should come as no surprise to find that fiction films, in recent years, have echoed a Deleuzian connection between brain and screen in the configuration of human memory. Whereas movies such as *Eternal Sunshine of the Spotless Mind* and *Memento,* described in Chapter 2, capitalize on confounding past and present in an individual's mind, science-fiction films such as *The Matrix* and *Strange Days* typically envision the potential of media technologies to mix past and future memories.[10] The advent of digital technologies, and particularly virtual reality environments, inspires the creation of virtual consciousness, allowing either a glimpse of the past through the eyes of a former self (*Strange Days*) or offering an embodied perspective on future events (*The Matrix*). In these movies, media technologies enable individuals to escape the constraints of the present, allowing them to move effortlessly between past, present, and future. Philosophical concepts articulated in science fiction films, in turn, help shape our mental and technological constructs of memory. The function of these movies is to "imagineer" what memory may look like in the future. A projected collapse of the brain with technology, far from an incidental concept, is a recurring trope in science fiction films.

One film that mingles a projection of memory's future with an imaginative design of the home movie as a conflation of brain and screen is Omar Naim's *The Final Cut* (2004). The movie is staged in the unspecified future, where the latest hit in home movie technology is the so-called Zoe Eye Tech Implant: an invisible organic device implanted in the brain of a fetus, equipping that recipient to shoot a lifetime of experiences through the eyes—an audiovisual recording straight from the brain. The implanted camera starts rolling the moment the fetus passes its mother's birth canal, and it is not removed until that person's death. In Naim's imagined society, one out of five people carry an implant, often unwittingly, as parents are not supposed to inform their children of this surprise until the children turn twenty-one. The Zoe implant is one of the most precious gifts a parent can donate to a child; it replaces the need for personal photographs or home movies because the eye camera offers a full and impeccable registration of someone's life. After the person's death, the chip is removed to be edited into a so-called rememorial, a ninety-minute film reminiscent of a made-in-Hollywood biopic. The implant comes with a new ritual: the "Re-Memory" is a commemorative gathering taking place forty days after a person's death at a cinema-alias-funeral parlor where the feature-length retrospective of the deceased is premiered to an invited audience.

The main character of *The Final Cut* is Alan Hakman (Robin Williams), one of the best professional cutter's of Re-Memory movies; his job is to turn the lifetime reels, removed from the deceased's brain, into conventional audio-visual productions, a sort of edited and anthologized digest presented as the ultimate obituary. In the main plot line, Hakman is asked to perform the "final cut" on the life movie of Charles Bannister, one of Eye Tech's attorneys who recently died. His widow Jennifer instructs Hakman to edit her husband's implant footage into a glossy retrospective, honoring the principles "family, community, career" and thus carefully omitting any scenes that would compromise his public image. As all cutters do, Hakman betrays the cutter's code proscribing to refrain from manipulation, instead accommodating the wishes of surviving relatives. While sorting through Bannister's life files, the editor is not only confronted with dubious money laundering schemes and fraud, but he also witnesses scenes in which the attorney commits adultery and incest with his daughter Isabel. Hakman discretely erases all compromising evidence from the deceased's ultimate portrait: the delete function turns out to be the cutter's most important Re-Memory tool. The result is a sanitized life in review, a public version of a family man with a brilliant career. Although every Re-Memory visitor understands the subtext of this genre, the dark side of Bannister's life remains invisible to the public eye. The ultimate home movie is everything but a true memory of the deceased's life. Like any film, this one is a mediated construction rather than a concatenation of pure registrations of authentic past events; after all, the editor defines both the choice and order of scenes. By definition, the retrospective captures the idealized family, and by removing painful episodes of adultery and incest, the family's memory is literally cleansed of its troubled past.

Director Naim takes Deleuze's idea of the brain as screen very literally: the eye is the camera, and the home movie is allegedly shot straight from the visual cortex. The theory of the movie camera coinciding with the mind's eye and the screen with the brain suggests the inseparability of thought processes from the technological substrates enabling their manifestation—what Deleuze touts as cinema's "psychomechanics."[11] The vector of time is an arrow bent into a circle: projections of the past become memories of the future, and vice versa. Naim's movie conveys a philosophical reflexivity based on Deleuze's contention that cinema is not about concepts but is itself a conceptual tool, raising questions such as: What are the new forces unleashed by memory once it becomes an organic biological-mechanical construct? What are the ethical and social consequences of eye implants?[12] Updated to Hansen's conjecture of digital technologies, Naim underscores the bodily basis of vision by making external and internal views indistinguishable. We constantly switch from the mind of Hakman, whose filtering of information is clearly tainted by his own memory, to images "shot" by Bannister's chip implant—the very images Hakman is supposed to filter. Instead of simply receiving images as pre-existing technical forms, mind-images and movie-images are mutually constitutive, consequently questioning the ultimate reliability of bio-engineered vision.

However, the convergence of brain and technology is not enough to account for this intricate construction of hindsight. *The Final Cut* attests to the idea that, despite the merger of eye and camera, the ultimate manipulating power of memory lies neither in the brain nor in the technology nor in a combination of both but in the interaction between brain, technology, and *culture*. With the help of advanced media technologies, every Re-Memory made up of mind-footage is modeled after the accepted cultural form of an audiovisual obituary, a life-in-review movie that is inevitably modeled after the conventional Hollywood format. In fact, the brain is the camera shooting the movie, but that movie is ultimately a product of culture. Capturing and projecting the family, even in the futuristic high-tech bio-digital society of science fiction, remains the work of an editor whose final cut is subject to the norms and social codes of the present. As becomes eminently clear from Naim's movie, mind and brain may fuse with technologies of memory, but all mental and technological constructs of past family life are always also social and cultural constructs.

What we learn from this film is that the construction of hindsight is arduously interwoven with dominant social and moral codes, anchoring futuristic extrapolations of techno-engineered memory in hegemonic cultural forms. The malleability of memories over time is not in the least facilitated by the technologies enabling their conception and later revision. In a way, technologies of memory facilitate the flexible interweaving of past, present, and future that our minds proffer. We wield camcorders and home videos to construct a pleasing future memory of our family's past life, thus anticipating the editing function as a feature of the mind as much as a feature of technology. And we wield film cameras to construct a version of memory that accounts for the conventional use of home recording technologies, while concurrently reflecting on the potential formative power of future technologies. At first sight, Naim's science fiction movie offers a straightforward illustration of Deleuze's theory of embodied memory as an amalgamation of mind and technology. Closer analysis yields the inescapable significance of cultural forms as constitutive elements in the construction of cinematic hindsight.

The "Home Mode" as a Techno-social Construct

Whereas Deleuze and Hansen focus on the merger of brain/mind with technology as the preeminent junction to study the meaning of moving images, cultural theorists find themselves more comfortable at the intersection of technology with sociocultural forces. In his excellent study of the home video, James Moran theorizes the historical and technological specificity of what he calls the "home mode"—the place of home movies or home videos in a gradually changing media landscape.[13] Rather than identifying the home movie or home video according to its ontological purity or as a technical apparatus, Moran rethinks the home mode as a historically changing effect of technological, social, and cultural determinations—a set of discursive codes that helps us negotiate the meaning of individuals in response to their shared social environment.

The home mode is not simply a technological device deployed in a private setting (the family), but it is defined by Moran as an active mode of media production representing everyday life: "a liminal space in which practitioners may explore and negotiate the competing demands of their public, communal, and private personal identities."[14] The home mode articulates a generational continuity over time, providing a format for communicating family legends and stories, yet it concurrently adapts to technological transformations, such as the introduction of new types of equipment: first the movie camera, later the video, and more recently the digital camcorder. Moreover, the home mode is affected by social transformations such as the position of family in Western society. Moran poignantly sums up: "While we use these media audiovisually to represent family relations to ourselves, we also use family relations discursively to represent these media to each other."[15] The changing depictions of families on public screens, most notably television, are part and parcel of the sociocultural transformation he is trying to sketch.

It is imperative to understand the evolution of the home mode and its technological and sociocultural constituents, because without recognizing these historical roots it will be difficult to account for its specificity as we enter the digital age. From the early beginnings of film, consumer technologies such as movie cameras have been drafted into the depiction of family life, whether as tools for idealization or as tools for inquiry and criticism. Although quite a few studies have been written on home movies in relation to the history of film or photography, few scholars have paid attention to the transformation of technologies in conjunction with changing social and cultural patterns of family life. Patricia Zimmerman explains, in her classic study of amateur film in the twentieth century, how the invention of increasingly lighter cameras seduced ever more parents into chronicling their children, thus providing a visual homage to the familialism of postwar America.[16] Zimmerman and other film historians tend to uncritically transpose the technological and ideological effects of the home movie as an instrument for promoting the ideal nuclear family in the 1950s onto home modes prevalent in later decades.[17] Moran, by contrast, convincingly argues that in subsequent decades the conventions of family representation changed in conjunction with home movie technologies.

The growth of the suburban family in the 1950s is inextricably intertwined with the emergence of television and of home movie cameras as domestic symbols of individual wealth and social cocooning. As television entered the private homes of the 1950s, images of screen families started to fill living rooms across America and shaped the concept of the nuclear family unit. Not coincidentally, many televised families, such as those portrayed in *Leave It to Beaver* or *The Adventures of Ozzie and Harriet*, conflated real life and screen reality.[18] Capturing one's own family in the 1950s and 1960s often meant to imitate the idealized family as shown on TV. The possession of an 8 mm camera in itself signaled the newly acquired material wealth that was prominently showed off both in television series and home movies of these decades. A parent's home movie camera functioned as a confirmation of intimate family life, an amateur production that defined itself against the increasingly popular public images of families on television; home recordings also served to entertain the family in their present. As objects of memory, home movies feature a family's life as a concatenation of ritual highlights, from birthday parties to first steps and from weddings to graduation ceremonies. The ability to record everyday events, to construct family life as it was, signified the individual consumer's power to model bliss and happiness after the ideal shown on television.

As the 8 mm movie cameras of the 1950s and 1960s gave way to the video cameras of the 1970s and 1980s, the style and content of the home mode changed accordingly, eclipsing the hegemonic portrayal of idealized families even if never replacing it.[19] In terms of its material apparatus, the light-weight video camera equipped the amateur user with an unobtrusive instrument to record everyday life. Video's ontology, unlike film, was based not in chemical but rather in electronic image processes, allowing for an unmediated display of moving images on the television screen. Video culture, as British media theorist Sean Cubitt contends, promoted the "metaphysics of presence": a documentary style that favored the inconspicuous presence of a camera as a fly on the wall, as if the filmmaker were part of the furniture.[20] It would be far too simple to reduce the vérité style of an era to the effect of a newly introduced media technology, but it is certainly no coincidence that video provided a way for capturing ordinary people's lives as they were in addition to the prevailing idealized way of recording them in the earlier home movies.

The home video became a favored instrument for recording quotidian reality, even if this reality did not live up to the traditions of family portraiture. A light-weight video camera lends itself much more easily than the 8 mm camera to unexpected and unobtrusive shootings; a family row or a sibling secretly stealing a cookie no longer evaded the discrete eye of the camera. And that new apparatus seemed particularly suitable to record everyday family life that was quickly changing in the wake of larger social and cultural transformations. A political climate with waxing protests against established norms of patriotism and paternalism defined this new generation of young adults; large numbers vocally opposed their parent's values about class, race, gender roles, sex, and ethnic identity. The nuclear family became a contested concept, as a new generation paved the way for sexual liberation, emancipation, and social diversity.[21] Home video defined itself in opposition to home movies as a mode subversive to the nuclear family ideal epitomized by 1950s home movies. In addition, video also came to define itself in opposition to television as a tool for rebelling against mainstream, collective values: community activists and gay and women's liberation movements used this medium to spread alternative concepts of identity and lifestyle to counter homogeneous mass media images.

The shift from idealized families and home movies to real families and home video also prominently figures in the production of domestic television series in the 1970s and 1980s. Images of family conveyed by typical postwar television series, and prolonged by series such as *The Brady Bunch* and *The Partridge Family*, were gradually supplemented by portrayals of less ideal and more realistic families, such as the Bunkers in Norman Lear's popular *All in the Family*, which first aired in 1971. Frequent confrontations between Archie Bunker and his son-in-law Michael ("Meathead") played out contemporary generational conflicts over political convictions, gender politics, and race relations in a society characterized by upheaval and rapidly changing norms. And yet, television never single-authored the hegemonic identity of the American family, just as the home video never uniquely covered the private lives of relatives. Television and home video mutually shaped the home mode in this historical time frame.[22]

There is no better example to illustrate the entanglement of television and home video in this era than the PBS series *An American Family*. A new genre, namely, family portrait documentary, was coming of age in the early 1970s when young filmmakers started to experiment with the new techniques of cinema vérité.[23] *An American Family*, a twelve-part documentary series that premiered on January 11, 1973, captured seven months in the life of a real California family—Pat and Bill Loud and their five teenaged children, Lance, Kevin, Grant, Delilah, and Michele. Producer Craig Gilbert and filmmakers Alan and Susan Raymond followed each member of the household at a turbulent time of their lives: the Louds' marriage ended in a divorce; the oldest son, Lance, announced he was gay; and the family unit was literally splitting up. A PBS announcement described this new television format as it invited viewers to "meet TV's first real family tonight and share their lives in the 11 weeks to follow."[24] The immensely popular series struck a chord with the American audience, as the Louds' turmoil mirrored quite a few experiences that real families encountered in this decade of change. The center was not holding—to echo Yeats's famous dictum—and television was recording its falling apart.

Because the series itself, as a unique and revolutionary media phenomenon, has received due scholarly attention, here I concentrate on the construction of the home mode in *An American Family*, because it exemplifies the struggle with codes inscribed in the home movies of the 1950s in favor of the newer, more immediate style of home video and cinema vérité.[25] *An American Family* subscribed to this new style as it suited the function of capturing family life as it is. In the words of the PBS announcement: "The Louds are not actors. They had no scripts. They simply lived. And were filmed."[26] Film crews that followed the Louds made themselves unobtrusive and their presence went almost unnoticed by the Loud family members, even while recording the couple's rows over Bill's extramarital affairs and their decision to file for divorce. Lance, the oldest son living in New York, received a light-weight Portapak recorder to document his own observations in addition to being filmed. This direct cinema style sharply contrasted the nostalgic photographs and home movie reels edited into the series as a way to visualize the Louds' past family life. As Ruoff explains in his reconstruction of the series: "The home movies and family photographs themselves represent an important detour from the observational focus on images and sounds recorded in the present. The Louds' home movies chronicle festive occasions such as Thanksgiving dinners and baby Lance taking his first steps. These nostalgic recollections suggest happier times, offering a powerful contrast to the Louds' contemporary lives."[27] Those incremental snippets of home movies and photographs not only signify life as it was for the Louds, but they also serve as a confirmation of collective memory of the 1950s as an era of happiness and peaceful family lives supported by ceremonial, ritual signposts and cemented in a transparent social structure with distinct roles for each family member. The camera registering life as it is stands in opposition to the previous home mode in both style and content, as it articulates the era's crumbling normative ideals by fixating the camera on the other side of family life: an emancipating housewife, a homosexual son, an unfaithful husband. The Louds' new mediated memories are recorded in a different fashion—using video and light-weight cameras—because the family can no longer live up to the ideal that was previously constructed through the lens of the home movies.

The oppositional home modes featured in *An American Family* are clearly in dialogue with its fictional counterparts on television, both conventional and more subversive sitcoms such as *All in the Family*. But whereas the fictional generational conflicts played out by the Bunkers are filled with explicit conflicts concerning sexuality, politics (Vietnam), identity, and race, the confrontations in America's "first real family on television" are much less ideological, only casually alluding to pressing political events or social debates raging at that time. Nevertheless, the realist record of a family falling apart, even if this reality was presented in the condensed and edited format of an eleven-part series, became a milestone in the history of American television. What attracted the audience at the time was most likely the way in which the Louds explored and negotiated the competing demands of their personal identities and prevailing family values in front of the public eye. The documentary camera, long before the introduction of so-called reality TV in the 1990s, became a constitutive element in the shaping of family life and personal identity.[28] Capturing the family, as a scheme for understanding and remembering, became integral to family life as a means to negotiate notions of individuality and togetherness, of deviation and belonging.

As poignantly put forward by Moran, every new media apparatus affects practices of production in conjunction to ideologies of home that reconstitute the family as a discursive domestic space. Now, if we extend this thesis to the last two decades—an era in which camcorders and digital equipment came into vogue—how does the home mode change along with notions of screened families? In the digital era, screened family life is paradigmatically defined by the ubiquitous presence of surveillance cameras and privately operated webcams. Whereas fly-on-the-wall cinema techniques set the standards for arresting reality in the era of analog video, the webcam may currently count as the symbolic catching hook for life as it is. A naturalistic mode of filming gives way to a surveillance mode of recording: fixed webcams cover an unwitting subject's movements. Even when the subjects are aware of the cameras' presence, there is no actual intervention from a camera crew. This mode of surveillance is not only frequently employed in public areas, where cameras control the movements of passengers, but with the webcam installed in a home, it is also rapidly becoming a common means of self-exposure on the World Wide Web.[29]

The digital mode seems to better suit contemporary fractured notions of family and more prominent individualism. Extending the home video logic of the 1980s—the real family captured in their rebellion against normative domestic values—the camcorder of the 1990s allots even more power to individual users to construct their idiosyncratic views of family.[30] As Moran observes, families in the 1990s and the twentieth-first century are no longer "natural" units: they are "families we choose"—domestic relationships between individuals construed as family ties.[31] The family we choose is the family that chooses to film itself as a unit, using an insider's perspective and camera, giving each member a direct voice in his or her representation. New conventions of television programming also define what constitutes a real family: a number of individuals who voluntarily move into a house, succumb to a regime of created conditions, and are continuously monitored by numerous surveillance cameras. The *Big Brother* effect, in a way, is the televised, formatted counterpart of circuited webcams installed in a family's home, continuously beaming pictures of real family life on the Internet.

Let us look at one example of a turn-of-the-millennium webcast featuring a regular family and compare it with a typical television series integrating the surveillance home mode into its program format. The webcam site of the Jacobs family from Alta Loma, a Los Angeles suburb, is one of thousands of sites featuring regular family life on the Internet these days.[32] Consisting of two parents and seven children, the Jacobs keep up an extensive home page where each member presents him- or herself in pictures and texts. The site presents an interesting mixture of the 1950s idealized family–style pictures and the digital surveillance home mode. When turned on, the webcam beams live footage from the living room: the children joking, the mother showing things to the webcam, the father leaning backward in his office chair.[33] The webcam footage seems to verify the real happiness professed in the website's texts and still pictures, as if saying, "Look, this happiness is authentic; you can watch us live from our home, every day."

Interestingly, the Jacobs family is literally made possible through the Internet. On their home page, they introduce themselves as follows: "We are the Jacobs Family. Clarence and DeShawn were married on November 27, 1999. We met on the internet [*sic*] in August of 1999, fell in love, and were married in Las Vegas. When me met, DeShawn had her son Dustin, and Clarence had Jennifer, CJ, Kyle, and Davey. Since then, we have 'added' Jarrett and Julia. A true 'Brady Bunch' as we are sometimes called."[34] The Jacobs, as we can see and read, are a typical postmodern fractured family: DeShawn writes about her birth mother and the search for her biological father, daughter Jennifer lives with her mother and stepfather in Arizona, and Clarence's sons are getting used to their new half brother and half sister. We can safely assume digital technology not only facilitates but actually enables family life for the Jacobs: without the webcam, their daughter Jenny in Arizona could not participate in everyday life, and without the Internet, DeShawn would not have been able to contact her biological father. In short, the Jacobs construct their "bunch" through digital media, while at the same time (and by the same means) they also beam this family's domestic bliss to a potentially worldwide audience.

Perhaps not coincidentally, the Jacobses' favorite pastime, as we can read on their site, is watching reality series on television. One of the more popular domestic reality series featuring in the early twenty-first century is *The Osbournes*. Ozzie Osbourne, once the notorious band leader of the heavy metal group Black Sabbath; his wife and manager Sharon; and their teenage children Kelly and Jack, according to MTVs announcement, allow surveillance cameras to continuously peek into their private lives, and that footage is turned into "an addictive new docu-series."[35] Even if the announcement promises the series to be a continuous window onto the lives of a famed rock star and his family, the net result is a traditional television format. On the one hand, *The Osbournes* deploys technological devices to create a sense of reality: furtive handicam footage is alternated by fixed steadicam shots, which are occasionally used by a family member to stage an intimate tête-à-tête with the television audience. On the other hand, the MTV docu-series is edited into an episodic format that undeniably reminds us of conventional genres—in this case a hybrid of a domestic sitcom and a video clip.[36] The docu-reels in the television series bring the traditional format of the domestic sitcom (and traditional family values) up to the current technical standards of reality capturing.[37] Naturally, the home mode and family concept featured in reality series like *The Osbournes* inspire many family webcam sites.

Following the social constructivist logic of James Moran, audiovisual representations of the Bunkers, the Louds, the Osbournes, and the Jacobs, reveal the interlocking of home movie technologies and cultural forms such as (reality) television series. In contrast to Deleuze and Hansen, Moran considers filmed family portraits to be technical substrates interwoven with sociocultural norms and conventions. While (cognitive) philosophers show little interest in the sociocultural component of converging brains-cum-apparatuses, cultural theorists such as Moran tend to disregard mental-cognitive functions when describing the home mode. And yet, I think we need a merger of both approaches in order to fully comprehend the intricateness of current filmic constructions of hindsight, especially now that digital technologies prompt us to develop a renewed awareness of mediated memory as a fabric woven of cognitive, technological, and cultural threads.

Capturing a Family's Past in the Digital Age

As stated earlier, our remembrance of family is prone to constant revision; home movies or videos shape and feed our memories, as filmic reconstructions insipidly coil with mental projections of family. In the digital age, it becomes increasingly easy to refurbish old footage into technically smooth productions and revivify former memories while retroactively adjusting them to fit our present knowledge and norms. Indeed, digital video in many ways destabilizes the supposed naturalness of analog video, as an emerging digital infrastructure shifts the center of gravity from simply shooting to complete processing and from image-sound recordings to multimedia productions. Today's computer hardware and software allows for affordable near-professional standards of editing and full-fledged productions, complete with subtitles, sound, and sophisticated montage, all in our private homes. Burned onto a DVD, the family's summer vacation in Cuba is now an audio-visual product that may compete technically and stylistically with travel programs featured on television. In addition, traditional image-sound recordings of home movies and videos increasingly yield to multimedia productions integrating documents, pictures, texts, (moving) images, and links to webpages. Multimedia productions on DVD no longer privilege the chronologically ordered visual narrative prescribing a viewer's reading but promote browsing through a library of connected files and (sub)texts. The digital cultural form breaks with inscribed codes of sequential episodes, allowing past and present images—even if shot in different technical modes—to merge in a smooth media product.

New technological devices, such as the digital camcorder, the World Wide Web, the webcam, the DVD, and the compact disc are of course not in and of themselves triggers for new mental concepts. A medium is both a material and a social construct, whose metaphors and models provide a horizon for decoding present knowledge.[38] Digital tools appear to give individuals more autonomy over their (multi)mediated portrayals of a family's past: mentally, by projecting remembered family history onto a new media product; technically, by editing old footage into reconstructed versions of family life; and culturally, by weaving in fragments of public footage (such as newsreels) with private portrayals. The emergence of a digital type of home mode does not make the filmic discourse of remembered family life more self-evident. Contrarily, because of their versatile and manipulative nature, moving images may easily become part of a dispute over disparaging versions of what family life was like. In watching contemporary (digital) reconstructions of family life, we need to account for at least three different levels of analytical awareness: First, we need to acknowledge how camera perspective and editing always imply an "I"—an individual imposing a particular view on a family's past. Second, we need to distinguish between various time levels implied in historical home modes; movie or video reels shot in previous eras beget a new illocutionary force when integrated in digital productions. And third, we need to remember the social codes and cultural contexts of historical and contemporary home modes while watching a production of screened family life. In the following analysis of a contemporary documentary, I demonstrate the significance of each level.

The documentary *Capturing the Friedmans* (2003) confronts viewers with the powerful role of various home mode technologies in authenticating and reshaping family life; director Andrew Jarecki's production is not just a documentary; the DVD-version, which I focus on here, renders a much more comprehensive view of the harrowing family saga he is trying to tell.[39] The Friedmans are a typical middle-class Jewish American family living in Great Neck, Long Island, where Arnold and Elaine, both into their fifties, have raised three sons: David, Seth, and Jesse. A retired schoolteacher, Arnold teaches computer classes for kids in his home basement and is helped by his youngest son Jesse, then eighteen. In November 1987 they are both arrested on charges of repeated sexual abuse of boys who attended their classes. The police raid heralds months of denigrating incarceration, release on bail, a media frenzy, a neighborhood witch-hunt, and—within the Friedmans' home—family rows over the best strategy to keep Arnold and Jesse out of jail. The family is torn apart by conflicting emotions of guilt, doubt, suspicion, and loyalty; David chooses the side of his father and brother and resents his mother Elaine, who is in more than one way the family's outsider. She is never convinced of Arnold's innocence, as he has lied to her in the past about his pedophilic inclinations and molestations. When her endearing attempts to save Jesse, by urging both father and son to plead guilty, inadvertently backfire, her oldest son David bitterly turns against her. In separate hearings, Arnold and later his youngest son enter their guilty pleas and are sentenced to substantial jail time; Arnold commits suicide in 1995 while imprisoned, and Jesse is released in 2001 after having served thirteen years of his sentence.

Although the events unfolding in retrospect before the viewers' eyes are dramatic by themselves, it is the hypertextual nature of this documentary and particularly the DVD that prevents the story from turning either sensationalist or partisan. Rather than following a chronological narrative logic, the documentary relies on the viewer's ability to identify three different technical types of film and to intertwine the distinct historical and contemporary home modes to which they refer: the home movies shot by Arnold in the 1950s through 1970s; David's home video footage recorded after the arrest in 1987; and interviews taped in the present by Jarecki.

For starters, the documentary features home movies and family pictures shot by Arnold Friedman primarily in the 1950s through the 1970s. These images are in perfect accordance with the conventions of home movies at the time: happy scenes of birthday parties, beach fun, and family vacations. Filming appears to be a family tradition—a narcissistic way of preserving the Friedmans' heritage on tape—as exemplified by an early 1940s recording that Grandfather Friedman made of his six-year-old daughter, performing as a ballerina in front of the camera. We learn from Elaine's voice-over that, several months after shooting this film, Arnold's sister died of lead poisoning. The joyful pictures of Arnold's childhood sharply contrast Elaine's commentary that her husband admitted to having repeatedly raped his younger brother Howard while sleeping in the same bed with him. Howard, interviewed in the present by Jarecki, desperately denies any recollection of his brother's self-confessed acts ("There is nothing there"). This incongruity between the ideal family life featured in the home movies and remembered reality cues the viewer to be suspicious when David Friedman, in turn, also claims to have nothing but rosy memories of his youth. His memory is corroborated by Arnold's home movies of his three sons playing in harmony with their father, having fun, and joking among themselves. Clearly, the home movies authenticate idyllic family life, but due to the comments by various family members, the viewer can only doubt their status as verification documents.

The second type of authentic footage stems from David's deployment of the video camera. Honoring the family's tradition, David had just bought a video camera in 1987, when the family started to fall apart following the arrest and the subsequent trial. In line with the conventions of cinema vérité and home video of the 1970s and 1980s, the camera keeps rolling as siblings engage in heated disputes at the dining table.[40] Mother Elaine often begs to turn off the camera, but the men clearly hold sway over the camera and ignore her requests. David and Jesse take turns in filming family rows but also record remarkable moments of frivolous acting—a sense of humor that obviously binds father and sons. The video camera, evidently deployed to capture life as it is, turns out to be just as unreliable as the old home movie camera capturing life as it was. Both home modes record a version of reality that later paradoxically serves as a desired benchmark for truth—whether this truth is a memory of ideal family life or a memory of a family at the verge of total disintegration because of false allegations.

The third type of footage, on-camera interviews conducted by Jarecki, reframes and unsettles the documentary evidence offered by pieces from the Friedmans' family archive.[41] Interviews with family members (Arnold's brother Howard; Arnold's wife Elaine and their sons David and Jesse, but not their son Seth, who declined to be interviewed) are supplemented by a number of interviews with people who were at that time involved in the Friedmans' indictment: their former lawyers, police investigators, alleged victims (then children, now adults, who both confirm and deny former allegations), parents of alleged victims, and an investigative reporter who wrote on the case. Mixing contemporary interviews with old news footage and trial tapes, the filmmakers manage to demonstrate the many angles on this case without ever privileging a single truth.

A condition for understanding the compilation of private reels is that viewers recognize the various perspectives at work in this production, laid in there by various family members who each try to steer and influence the outcome of this memory product and thus define the meaning of what happened to their family. The conundrum of slippery and quivering truth, which is clearly present in the screen version of *Capturing the Friedmans*, is even more palpable in the DVD version of the documentary. Needless to say, digital equipment was instrumental in the seamless cross-editing of the three (historical) types of recordings and in braiding together the various individual perspectives of the family members. Turning the movie into the cultural format of a DVD—often including footage of the making of scenes and evidentiary material—viewers assume the position of active co-constructors of the story. The DVD includes many extras: full interviews with witnesses for the prosecution, a family scrapbook with photos, more home video footage, and news reels on the case. In addition, a second DVD disc features coverage of the discussion after the New York premiere, where many people involved in the case dispute each other's versions of what happened to members of the Friedman family. And most significantly, the disc contains a section where the viewer can read key documents, such as letters from Arnold and a police inventory from a search of the Friedmans' house. As the extra documents become an integrated part of the puzzle, the viewer is encouraged to sharpen his or her judgment by reading more evidence, both to buttress the judge's decision and to back up the family's defense. In fact, as director Jarecki suggests in an interview, the documentary serves as "the trial that never was" (there were only hearings in front of a judge), and the audience serves as a jury. From the puzzle of recordings, viewers ultimately decide for themselves what happened to this family. The seamless web of digitized documents weaves the family's narrative into an open-ended hypertext of possibilities: facts, testimonies, truths, and illusions.

And the documentary is extended on the Internet via a website that updates its viewers on the continuous saga of the Friedman family. David and his brother Jesse—now released from prison—try to get a retrial, on the basis of Jarecki's film (especially his interviews with witnesses), to prove their father's and Jesse's innocence.

In line with Moran's constructivist theory, technical and socio-cultural codes codetermine the construction of family and our personal and collective memory of it. Yet beyond this constructivist analysis, we need to acknowledge that *Capturing the Friedmans* is also a coproduction of mind and technology, involving both the reconstruction of past images and the projection of future memories. Future construction of cinematic hindsight is already inscribed at the moment of each home movie shooting, most notably when David Friedman, in the middle of the turmoil in 1988, takes up the video camera and turns it onto himself. Sitting on his bed, and starts a monologue: "This is a private thing, you know . . . if you're not me, you're not supposed to be watching this. This is between me and me, between me now and me in the future." In this scene, David is addressing himself in the future. Indeed, why make a video if there was no intention of telling the family story someday, in some form? Another instance illuminating the intentional inflection of future memories of the family's past comes in David's response when director Jarecki asked him why he started to film his family's ordeal: "Maybe I shot the video tape so I wouldn't have to remember it myself. It's a possibility. Because I don't really remember it outside the tape. Like your parents take pictures of you but you don't remember being there but just the photographs hanging on the wall." David cogently identifies the power of home video and home movies as dual instruments for constructing and remembering family life. On the one hand, he needs to record his own version of reality because his father is going to jail and he does not want his own future children to remember their grandfather from newspaper pictures. On the other hand, he wants to document his father's and brother's innocence. For instance, David films Jesse while driving the car on his way to the courthouse where Jesse intends to enter his guilty plea and hopes to obtain a reduced prison sentence; David forces his brother to state his innocence when he asks, "Did you do it, Jess?" to which Jesse solemnly responds, "I never touched a kid." This home video footage painfully contrasts the official court video later in the documentary, showing a crying, remorseful Jesse admitting his guilt to the judge—an act so convincing you no longer know which documentary evidence to believe. David and Jesse undoubtedly utilize the home video to assert some measure of control over the events as they are unfolding, perhaps in an attempt to avoid the family breaking up. But in doing so, they consciously build their future defense—their personal memory of a torment that was, in their version, uncalled for and unjust.

As we learn from *Capturing the Friedmans*, family is a production of media as well as a product of memory, involving projection and reconstruction.[42] Through our home modes, we construct the memories of tomorrow, and via our technologies, we create projections of the past. Family is remembered through mental, technological, and cultural means of mediation, and every generation chooses its own tools to understand and reframe that concept. If we want to understand how both media and memory change along with notions of family, we may depart from Jarecki's improvised definition of memory in an interview included in the DVD: "We think we can put our memories away in a box and we can go check on them later and they will be the same, but they are never the same; they are these electro-chemical bubbles that continue to bubble over time." Memories never just are; they are always in the process of becoming. Our memories of family evolve in conjunction with home movie technologies and social codes and conventions; the real challenge of cultural analysis is to recognize mediated memories' historical complexity in order to explain their contemporary relevance.

The Future of Home Movies

The home mode, besides referring to a screen registration and a social construct, also involves a mental projection of family life. Remembered families are hence projected families—the simultaneous products of mind and matter, of home and Hollywood. Therefore, the future of memory will be determined as much by our tools for remembering as by our imagination. Cultural forms—such as television series, documentaries, family websites, science fiction movies—and the technical tools by which they are produced, intimate from the minds that peruse and deploy them. Deleuze states that cinema makes the invisible perceivable: in movies, past reconstructions and future projections materialize into image sequences, which in turn feed the viewer's imagination. Cinema is a matter of "neurophysiological vibrations" where the image "must produce a shock, a nerve wave which gives rise to thought."[43] Memory moves along the axis of time: we are always in a state of becoming. The cultural forms we produce, whether home movies or science fiction films, are at once the result of, and input for, our brain waves. The future of home movies is contingent on the shaping power of past and present (filmic) productions—a shaping power that leads Hansen to shift his focus to the "post-cinematic problem of framing information in order to create (embodied) digital images."[44]

That same circularity applies to the model's horizontal axis relating self to others. Individual family members produce and project home modes while the home mode defines the concept of family as a social unit of belonging. What we call a home movie is in fact a coproduction in which mental and cultural concepts of family and home constantly evolve, at once revising old notions and anticipating new ones. Changing technologies (8 mm film, video, camcorders) are instrumental in the construction of familial memories—images of how a family was, how it presents itself, and how it wants to remember in the future. But the filmic perspective on family is always embodied in individual minds: individual members whose subjective pairs of eyes provide points of view that may at any time diverge or converge from siblings or other relatives. A home movie, like memory itself, is not a self-evident filmic document that chronicles a family's past. Instead, by analyzing the discrepancies and tensions between its various makers and producers, we acknowledge moving images stored in our family archives to be input for—rather than output of—memory acts. The meaning and impact of these documents may always be subject to future reuse and reinterpretation, and hence our notion of family is never established for once and for all.

With the advent of camcorders and advanced editing facilities on personal computers, the awareness of moving images serving as input for future memories is likely to become more prevalent. The rapidly growing cultural practice to record one's life audiovisually by means of ever more digital instruments, combined with the innate inclination of human memory to select and reinterpret the past, presages the immanent expansion of cinematic retrospection. Bolstering this trend is the growing interest of people in multimedia productions that galvanize their remembrance after death; like artists or actors, we want to secure an eternal place in the virtual universe.[45] Personal recordings of someone's life increasingly resemble fashionable television formats or conventional film genres. A perfect example is the popularity of memorial videos as part of a funeral experience. Businesses such as Life on Tape and Precious Memories offer the possibility to turn pictures and home video footage into a smooth five-minute eulogy to be screened during the memorial service or to be burned on a DVD as a gift to family and friends after the funeral.[46] Reminiscent of the fictive re-memorial in *The Final Cut*, the five-minute eulogy proffers a seamless blend of personal (moving) images into a standard format of pre-selected soft-focused imagery, complemented by the deceased's music of choice. A recent trend to shoot and edit your own memorial movie while still alive and edit the footage into a memorial video—which not coincidentally resembles the biopics of public figures broadcast on television upon their death—seems the next stage in the construction of cinematic hindsight.

Mental projections, technical imagineering, and cultural imaginations can hardly be analyzed as separate manifestations of audiovisual remembrance. Therefore, we need both Deleuze's and Hansen's concepts exploring the construction of memory at the junction of mind and technology, as well as Moran's constructivist theories analyzing the home mode at the intersection of technology and culture. Cinematic constructions of family-life-in-review are the result of concerted efforts to save and shape our private pasts in a way that befits our publicly formatted present and that steers our projected futures. Combining (cognitive) philosophical with social constructivist perspectives and cultural theory, we will be better equipped to understand future constellations of home modes as the multifarious products of mind, technology, and culture.

7

From Shoebox to Digital
Memory Machine

An old friend recently admitted, with a sense of understatement, that the size of his personal digital collection has outpaced his ability to keep track of its contents. Since acquiring a digital photo camera and a scanner in 1996, he has taken, stored, and scanned well over a hundred thousand pictures of his daily life, work, and family. His collection of DVDs and audio files also faces the fate of infinite expansion due to the increasing availability of downloads. Combined with occasional camcorder activity and heavy Internet use, the act of recording and storing files, images, audio, and data absorbs nearly all of his spare time. Space is no longer an actual or virtual constraint, because only my friend's prerecorded movies on DVDs are still stored as material artifacts on the shelves, and computer RAM has become relatively inexpensive. His proposal to transfer the family's entire collection of old photographs and videotapes onto digital media met resistance from his partner who is attached to the touch and feel of analog products. In addition, digitization confronted my friend with issues of time and order: time to enjoy and relive recorded cultural and personal moments, while constantly engaged in capturing and storing the newest and latest experience; and order to allow the retrieval of specific moments, as the danger of getting lost in his multimedial repository was growing by the day.

Previous chapters illustrated how the gradual takeover of analog by digital technologies has impacted various forms of personal cultural memory; this chapter focuses on the accumulation, storage, and retrieval of digital recordings. Multimedia technologies offer new opportunities in the everyday lives of people but also impose new complexities; with ever-expanding memory capacities, the computer seems to become a giant storage and processing facility for recording and retrieving bits of life. As more people enjoy the pleasures of digital equipment, many of them, like my friend, also acknowledge the problems that come with new technology, such as handling exploding quantities of personal data. Although we tend to attribute both the bliss and dilemmas of expanding memory collections to digitization, they are neither completely new nor uniquely related to the computer age.

History has given rise to a number of fantasies imagining all-encompassing solutions to the fallibility of human memory. Since early modernity, people have tried to imagine and invent memory machines that could remedy two basic shortcomings of the human brain: its inability to systematically record and store every single experience in our lives, as well as the brain's incapacity to retrieve these experiences unchanged at any later moment in time. Fantasies such as Vannevar Bush's memex are based on such mechanical notion of memory, aiming for ubiquitous recording and accurate retrieval of recorded documents.

Historical blueprints of memory machines were premised on static models of human memory, so it will be interesting to see whether contemporary designs account for more recent insights in the dynamic nature of memory. Commercial ventures are quick to provide digital alternatives to familiar analog forms, such as digital photo albums, weblogs, and audio software, facilitating the storage and retrieval of personal files on the computer. Understandably, marketers tend to focus on digital products that assume memory to be a locus of retention—a library or archive. Software engineers, working for both commercial developers and public research institutions, have recently started to address the question of comprehensive memory storage by designing digital tools that may accommodate large quantities and varieties of personal digital files. Four such projects are analyzed in more detail in this chapter. Whereas some projects simply promise to solve the shoebox problem, others aim at designing completely new systems of memory storage and retrieval; yet others predict that their software and hardware will revolutionize our very ability to remember. Despite their bold claims, it is surprising to find how most projects adhere to the mechanical logic espoused by Bush and regard the brain and the machine (memory and media) to be ontologically distinct entities.

Because a main thesis of this book is to show the mutual shaping of human cognitive memory and media technologies in everyday cultural contexts, I address the questions of how new digital memory machines may affect the nature of our recollections and the process of remembering and, vice versa, how a connectionist model of human memory and its recycling mechanism influence the conceptualization of new tools. The last section of this chapter argues that networked computers serve less as repositories and more as agents of change. Digital storage-retrieval facilities, such as search engines, are not merely new metaphors that mold our concepts of memory; they actually define the performative nature of memory. Identifying three major types of transformation—digitization, multimediatization and googlization—I argue these developments are integrated in the ways we store, retrieve, and adjust memories in the course of living.

Fantasies of a Universal Memory Machine

Fantasies of the perfect memory machine—a machine that stores everything and keeps its items systematically ordered in immaculate condition—have always accompanied the invention of new media tools; the age of modernity gave rise to a number of technologies heralding broader social and cultural transformations. As Australian media sociologist Scott McQuire points out, the split of living memories from so-called artificial or technological memory, first made possible by the invention of writing, engendered dreams of complete recordings as well as systematic orderings and the retrieval of lived experiences.[1] Both the German philosopher-mathematician Gottfried W. Leibnitz and the English mathematician Charles Babbage are credited with visions of mechanized memory tools, which in hindsight are seen as early precursors of the computer.[2] Yet the most famous visionary of the modern memory machine is undoubtedly Vannevar Bush, former director of the American Office of Scientific Research and Development, whose fantasies had more than a little impact on the ideas of contemporary engineers and scientific communities.[3]

In his famous essay "As We May Think," published in 1945, Bush expresses his fear that society will soon be bogged down by an explosive growth of specialized publications, and he urges scientists and engineers to turn to the massive task of making our bewildering stock of knowledge more orderly and accessible.[4] Placing himself in the tradition of Leibnitz and Babbage, who both envisioned extensions of the mind in the form of calculating and arithmetical machines, Bush commits himself to designing a memory machine that enables the storage and retrieval of various types of records: documents, photographs, film, television programs, and sound and speech recordings. His essay contains descriptions of imagined recording devices, including the Cyclops Camera ("Worn on forehead, it would photograph anything you see and want to record. Film would be developed at once by dry photography") and a vocoder ("a machine which could type when talked to").[5] The first invention is reminiscent of the Zoe Eye Tech Implant, the science-fiction device described in Chapter 6, whereas the vocoder is now a successful commercial product also known as speech recognition software. But Bush's essay is most famous for the prediction of a new type of machine allowing humankind to avoid repetitive memory tasks and thus make room for more creative thought. Analogous to the idea of calculating machines, Bush suggests the memex: "a device in which an individual stores all his books, records, communications, and which is mechanized so that it may be consulted with exceeding speed and flexibility."[6] An essential feature of the memex is its ability to automatically select and retrieve every stored item swiftly and efficiently.

Bush's memex fantasy has been hailed as the greatest vision in anticipation of the computer age, but it has also been criticized for its ideological undercurrents of "pioneerism" and "frontierism."[7] For the purpose of my argument, however, I am less concerned with Bush's general ideas about science and progress and more interested in his assumptions about the relationship between the human brain and the memory machine. He proposes to model his memex device after the human brain in order to artificially duplicate the mental process of memory retrieval, thus relieving the brain from a number of repetitive tasks. Conventional storage and retrieval systems, which classify data alphabetically or numerically and in which information is found by tracking it down through subclasses and indexing, are cumbersome and counterintuitive. According to Bush, the human mind operates by association and so should memory machines. Note that the mind is both a model and a metaphor for the machine—two separate entities connected figuratively. In the history of information technology up to the digital age, the brain has been frequently invoked as a source of metaphorical imagery, mirroring the frequent use of (media) technologies in neurobiological and cognitive circles, as described in the previous chapter.[8]

Bush's concept of the memory machine is both mechanical and paradoxical. Mechanical because he presumes an unambiguous vector between technology and the human mind: the memex ought to function as a human mind. Memory, to his regret, is fallible ("transitory") and therefore, a machine should take over part of the brain's function to prevent amnesia due to information overload. Ideas, memories, and thoughts are stored in documents or other recordings, and these recordings should be randomly ("associatively") retrievable. When talking about data, Bush equates bits of information to ideas, memories, and thoughts that can be put away in a repository and be pulled out in random order. However, retrieval of documents from a database and retrieval of memories from a human brain are fundamentally different processes with very distinctive goals. Documents or recordings can be stored in a database, and we want them to be there, unchanged, as we retrieve them and subject them to (re)interpretation; memories are never unchanging data that can be stored and retrieved in original shape. As German media theorist Hartmut Winkler puts it: "Material storage devices are supposed to preserve their contents faithfully. Human memories, on the other hand, tend to select, reconfigure, and forget their contents—and we know from theory that this is the real achievement of human memory. Forgetting, in that sense, is not a defect, but an absolute necessary form of protection."[9] Even if human memory and material information deposits are distinctly different entities, they are inextricably intertwined in the process of remembering: retrieved documents constantly feed upon a twisting and changing memory, whereas the human mind tends to alter information in a depository by replacing, renaming, or deleting its content files. As most people know from experience, reordering shoebox contents or reorganizing one's files on a desktop always involves both a mental and a material exercise. Of course, deleting or destroying items may cause intense regret later, but that regret is part of the integral act of remembering one's past.

Returning to Bush's concept of memory, we can conclude that, besides being mechanical, it is also paradoxical. Machines cannot simply be modeled after the human brain, because the brain interacts with the machine and vice versa. In Bush's vision, there is no room for the logic that technology also structures, rather than simply reflects, cognition—a serious flaw that is echoed in quite a few contemporary hypertext theories.[10] Hypertext enthusiasts proclaim that interactive software programs erase mediation, liberate the writing subject, and empower writers who were formerly restricted by the constraints of linear narrative discourse. Like Bush, they regard the human mind as the model for the machine rather than in interaction with the machine. Counteracting the commonplace notion that hypertexts are more natural (hence better) because they are more like the human mind, I here defend the notion that technologies of inscription actually shape human cognition.

Bush's fantasy of the universal memory machine, the memex, has inspired many recent projects concerning the inscription, storage, and retrieval of personal memories or "bits of life." In its 1945 utopian form, Bush's concept prefigured the need for an exterior digitized memory with unlimited capacity; it anticipated the transformation of personal collections into multimedial compilations of images, text, and sound—technical tools to record and invoke memories; and most of all, the memex foreshadowed the need for automated retrieval systems as a consequence of exploding quantities of information. In sum, the memex fantasy supposedly contains every ingredient to solve the giant shoebox problem, and it is therefore hardly surprising to find many recent projects citing Bush's fantasy. His mechanical and paradoxical assumptions about the machine's verisimilitude to the brain are equally echoed in contemporary digital projects. Four specific contemporary projects for designing a digital memory machine, discussed below, illuminate either how the brain is envisioned as the functional model for the computer or how the computer serves as a model for the dynamics of human memory. And yet remarkably, neither of these innovative projects anticipates the co-evolution of brain and machine, let alone the even more complicating factor of sociocultural practices and forms as decisive factors in the construction of digital memory machines.

Contemporary Visions of Memory Machines: Four Projects

The past decade has given rise to various initiatives—both commercially and privately funded—to solve the complex management of digital personal collections. There have been many more initiatives in this area than I have room to describe here, so instead of a general overview, I focus on four projects in more detail: Shoebox, the Living Memory Box, Lifestreams, and MyLifeBits. Some of these projects are technical in nature, concentrating on the development of hardware and software tools that help tackle the management of personal data; others are more visionary, espousing a comprehensive perspective on the nature and structure of personal memory machines—more in line with Bush's historic effort. Despite the different goals and outcomes of these various projects, their common assumptions about the functioning of personal collections in relation to memory betray a peculiar desire to control and manage the human brain like a computer system or, vice versa, to model the machine after the brain.

Shoebox

"Shoebox" is the name of a digital photo management system developed by AT&T labs in Cambridge, England.[11] Its software package provides a range of browsing and searching facilities for the storage and retrieval of digital photographs, utilizing spoken and written annotations as well as a content-based image retrieval method.[12] The AT&T researchers specifically tested the usability of Shoebox as a retrieval system in the context of personal photo collections, involving audio and textual labels besides photographic images. Although they rarely talk about memory as such, their research is built on the assumption that personal photo collections primarily—if not solely—serve as reminiscing tools in people's everyday life. Perfectly ordered photo collections save the user time and annoyance: easy retrieval depends on logical storage and labeling.

For instance, the easiest way to index series of photographs would be to order them chronologically (when taken) or geographically (where taken). Yet the designers resist the notion that personal collections of photos are exclusively stored or retrieved by factual data. Shoebox tacitly subscribes to the idea that people remember a picture topically (the Tower of Pisa) or recall a series of pictures thematically (our vacation in Italy). Adding key words or short descriptions, users can also retrieve specific photos by combining objects and persons (Uncle Sam at the Eiffel Tower). Frequent users may vaguely remember the content or image shape of the picture, but they may have forgotten when or where it was taken and thus, where it is stored. Image-based indexing, in that case, may be a helpful retrieval tool. The outcome of the comparative trial is that audio or textual annotations tagged onto digital pictures by the user are a more effective means of automatic retrieval from large collections than image-based indexing.

Although Shoebox is a strictly technical project concerned with the automated labeling and retrieval of photographs, its ensuing software products are peculiarly grounded in mechanical assumptions about personal memory. Researchers seem to reduce cultural practices to technical tools, and they project the logistics of the machine onto cognitive and cultural processes without actually scrutinizing the latter. For instance, Shoebox forces users to add verbal "tags" or textual annotations to photos, preferably when loading images into the system. Technical considerations dictate the moment of tagging: it occurs not when the picture is made but when it is stored away in the system, because noise is reduced when the spoken annotations are recorded in a quiet and controlled environment. However, annotations impact memories depending on their moment of attachment. While taking the photograph, one may add superficial or intuitive descriptions; while processing the pictures into the system at a later moment, the image is already colored by memory and has probably become part of a narrative— comparable to putting a picture into a photo album and adding textual comments.[13] But a picture in an album has a different function than a picture in a shoebox. Both are building blocks for personal memories, yet whereas the album is formatted as a narrative, the shoebox is a deliberately unsorted collection. The annotation, rather than a fixed bit of factual information, is thus a flexible constituent in the recollection process.

In light of the assumed technological imperative projected onto the working of personal memory, the disappointing conclusion of the Shoebox designers that "users may not be willing to annotate images and may never even wish to perform a search" does not come as a surprise.[14] The mismatch between the tested technical tool and the chosen application may rest, to a large extent, on the conceptualization of personal memory as a mechanical process rather than a cognitive and cultural one. Designed as an automated solution to a complex mental-technical-cultural practice, Shoebox ignores both time and order as shaping factors in the memory process; the attachment of meaning changes with the various stages of taking pictures, identifying them, ordering them, turning them into a narrative, and remembering them at a later stage. Identifying labels or indexes may vary with time, and so does their meaning in the memory narrative.

The process of remembering, in other words, is infused with time and order—determinant factors in the continuous shaping and contextualizing of past experiences.

The Living Memory Box

Whereas Shoebox is primarily devised as a software tool for photo management, the Living Memory Box (LMB) advances its view from computerized storage systems to the more expansive level of a memory supporting activity; four researchers at the Georgia Institute of Technology envision how a digital support system may help "enhance the memory archiving experience of today's families."[15] The LMB encompasses a central storage display combined with a portable recording device connected through an innovative interface design that makes it look like a transparent shoebox with a screen on top. The researchers, who have tested the prototype on several focus groups, emphasize the importance of usability: they want to know how individuals and families actively engage in storing and retrieving precious memories and to recognize a person's concerns in savoring his or her future personal heritage. The result is a holistic design that addresses software and interface models after a careful reconsideration of user needs with regard to memory activities.

In contrast to Shoebox, a system basically restricted to the storage of photographs, the Living Memory Box supports scrapbook activities, encouraging the inclusion of a variety of objects. Because its design limits the capacities of material objects to be tucked away, the "box" should be leveraged metaphorically: every cherished object may be photographed and included in the digital archive. Hence, a child's crayon art finds its way into the LMB through pictures of the actual drawing, and the memory of an impressive theater play may be stored away by saving a picture of the performance along with a picture of the entrance tickets or flyers. Another conceptual difference with Shoebox is that the Georgia Tech researchers regard memory as an active and discursive process that entails more than just saving and retrieving (photographic) artifacts. Users of the LMB system consider making digital scrapbooks a "time of personal expression" that is therapeutic and/or pleasurable. Therefore, the interface of the box allows for natural interaction between technology and users, such as the inclusion of human voices along with images and text in composing memory narratives. More important than storing pictures, say the researchers, is "to help users complete more pleasurable and complex tasks," that is, telling stories about particular events and then linking them to "related memories through the simple interface."[16] Telling stories helps memory come to life, hence it turns the box into a living family archive.

Even though the LMB is a major improvement over the mechanical interface of photo management systems, it runs into some of the same paradoxes concerning spatial order and time. One returning problem with digital storage is the attenuation of memory items through digitization: the necessity to render every three-dimensional, physical object digital before storing it may diminish the user's appetite for the system. Many people, after all, want to hang on to the actual object (such as the crayon drawing) rather than save a picture of it. Digitization opens up new potential because of hardware's infinite storage capacity, and yet this storage is limited to virtual objects only. Peculiarly, one of the conclusions coming out of focus group research is that in order to be successful, the LMB should bring interaction away from the computer: users want their act of memory to be more than a cut-and-paste activity based on interaction with the screen. Another paradox is that researchers acknowledge the importance of integrating the factor of time into their design but fail to accommodate the dynamic and versatile nature of memory. The LMB still works from the assumption that memories are stored, retrieved, and perhaps reassembled or reused, but in essence they remain unchanged. Needless to say, the design caters to parents' anxiety that they are not saving enough or are choosing the wrong memories, but the obsession with complete storage strangely distracts from the great potential of digital systems to exploit the inherently dual nature of human memory to both store and revise, to simultaneously save and delete, and to function both like an archive and like a story-generating machine.[17] Apparently, the concept of memory as an accurate retrieval machine is still the dominant model after which contemporary visions are sculpted.

Lifestreams

Whereas Shoebox and the Living Memory Box offer designs of concrete software-supported memory machines, the next two projects are more in line with Bush's visionary scheme. Introducing their Lifestreams project, four researchers from the Department of Computer Science at Yale University define their modest goal: "to change the world, computationally speaking."[18] A lifestream is a time-ordered stream of documents that functions like a diary of an individual's electronic life; every document you create is stored in your lifestream, and so is every document you receive from other people. Lifestreams are comprehensive recordings of someone's activities, registering every communicative and expressive daily activity mediated by the computer. In the electronic world to come, according to these computer scientists, software handles a good deal of the overhead that comes with managing chunks of information. By "chunk" they mean any piece of data that would ordinarily be treated as a unit: a document, an e-mail message, a calendar item, a software Rolodex card, a fax image, or the transcript or screen image created by an executing application. The Yale engineers promise a lifelong tracking system that disposes of cumbersome hierarchical filing directories and instead privileges "chunks of life," which require no explicit storage management system by means of labels: "If you are a student, your transcripts, graded assignments, bulletins and schedules are stored on your stream. If you are a patient, your medical history is stored as a series of discrete chunks (lab reports, prescriptions, doctors' notes). Whoever you are, item number one in your Lifestream-of-the-future is probably your birth certificate."[19] The tacit assumption underlying this project is that you are what you document, and "documents are us." Life's experiences are inscribed through machines and digital files that verify each second of one's existence.

In more than one way, the Lifestream software design subscribes to Bush's vision of a computed world in which the brain serves as a model for the computer. The machine, like the mind, creates streams or substreams that can be suppressed, activated, or trashed but never lost. A chunk that has not been worked on for a while simply becomes hidden in someone's virtual past, resting in the repository unchanged, waiting to be retrieved. This theory of lifestreams inadvertently resembles the stream of consciousness and subconsciousness concepts underpinning Freud's theory: experiences are always somehow preserved in human memory, either consciously or unconsciously, only to be activated at a later moment of recall. Yet though the machine is modeled after the brain, the computer is also upheld as a model for how the brain ought to work. Inscribed in the software is the idea that strings of information are accessible anywhere, anytime, anyplace (ubiquitous availability) as well as the presumption that our personal "lifestreams" are always connected to others' digitized "lifestreams," allowing one to copy a chunk from someone else's digital stream (ubiquitous connectivity).

Lifestreams thus foregrounds ubiquitous availability and connectivity but disregards the dynamic nature of both documents and memory.

Although Bush's metaphor—the brain as a model for the computer—echoes in the Lifestreams project, the multimedia computer hardly seems to *affect* human memory.[20] The new digital memory machine, in which various types of media modalities (audio, video, text) have converged, seems a perfect instrument to naturally record one's multifaceted, multisensory, and associative experiences, but it never influences the way we remember these events. Life's past episodes are considered holistic multimedial experiences, the inscription, collection, and recollection of which are conceived as an infinite database that can be exhaustively recorded and endlessly replayed. It is interesting to notice how this metaphorical equation hinges on the same assumptions about ubiquitous recording and accurate retrieval as its historical precursor. A corollary to this metaphor is a denial of interaction between human and machine as mutually constitutive entities.

MyLifeBits

Whereas Lifestreams sprouts from the minds of computer scientists, MyLifeBits is an American commercial venture, but the two projects operate from remarkably similar premises. MyLifeBits was launched in 2002 by a group of Microsoft engineers at the MS Media Presence Lab in San Francisco led by Gordon Bell.[21] The project leaders work on a comprehensive software system and communicate their goals and mission to a broad audience.[22] The engineers at Microsoft aim to build multimedia tools that allow people to chronicle their lives' events and make them searchable, because memories deceive us: "Experiences get exaggerated, we muddle the timing of events, and simply forget stuff," says one of the project leaders. "What we want and need is a faithful memory, one that records and builds on the reliability of the PC."[23] Bell and his team consider memory's ability to forget and revise a weakness rather than a strength or necessity. In the many interviews Bell issued to the news media, he pitches MyLifeBits not only as the solution to the giant shoebox problem but as an organizer of life: everyday events will be fully recorded in text, images, and audio and stored orderly in a computer. Each item will be tagged by audio or text annotations—tags that may also be cross-linked. To test the program, Bell has downloaded all his personal bits of life, including his parents' photographs, onto a hard drive; MyLifeBits, he claims, is more than a memory, it's "an accurate surrogate brain," the realization of Bush's memex machine, which equally featured automated retrieval as its highest ambition. Bell imagines how, someday in the future, a compulsive recorder will call up a single day in his or her life and get an hour-by-hour breakdown of what he or she did, said, and saw. Besides chronological retrieval, one major advantage of MyLifeBits software is its ability to allow a "Google-like search on your life": to retrieve random memories by typing in a tag.

Unlike Shoebox, but much in accordance with the Living Memory Box, MyLifeBits software departs from the notion of stories or memory narratives as key-ingredients of the remembering process. The shoebox of digital items is viewed as a nonhierarchical repository of annotated data, out of which users construct a story every time they retrieve a single bit. Annotated stories may be browsed like the World Wide Web, that is, the user simply follows the links connecting one resource to another by means of keyword. Thus, memory is conceived as an associative journey through linked bits of life, which may subsequently be re-presented in any mediated form: as a Powerpoint presentation, a slide show, or a photo album. Unlike other systems heretofore described, the presentation of stories enabled by item retrieval constitutes the conceptual heart of this project. The most valuable inheritance to our children or grandchildren, Bell claims, is not a shoebox of assorted items but a selection and representation of annotated stories. "Your cinematic deathbed flashback will already be uploaded to your hard drive," one of his interviewers concludes.[24] In MyLifeBits, the idea of the computer as a model for the brain is extended from its storage and retrieval capacities to its presentation capacities. It is interesting to notice how Microsoft engineers construct the notion of life as a story, while simultaneously equating life stories with mediated formats. Personal memories cast in narrative, using images as material signposts, conceptually morph into preformatted media presentations—preferably copyrighted by Microsoft, of course.

MyLifeBits's design deftly reflects (and smartly caters to) two contemporary anxieties: worrying about managing one's life and worrying about amnesia. For an upscale Western audience, managing data has become an attractive metaphor for controlling life. To live an experience at a date and time of one's choosing—rather like a television program recorded on a VCR—takes some pressure off life's fast pace, regulated by the clock.

What could be more appealing to a contemporary user struggling with time constraints in an experience economy than the storage of events in mediatized, retrievable memories? The anxiety brought on by missing events as one's children grow up can be assuaged by the thought of a personal memory machine, enabling precious moments to be replayed at a time more convenient than the ever-demanding present. Experiences etched in dimensions of time hence become a timeless repository of reruns. On the other end of the spectrum, the prospect of harrowing memory disorders, such as Alzheimer's disease, feeds on the fear of amnesia. Complete storage of personal memory and collections should avoid the erasure of someone's unique identity. The anxiety of forgetting is implied in the desire not to be forgotten: as MyLifeBits insists, the most important beneficiaries of this software product are your descendants. Immortality through software cultivation appears to be an attractive prospect in which to invest.

Shoebox, the Living Memory Box, Lifestreams, and MyLifeBits—projects that envision the future of memory machines—all capitalize on the digital enhancement of limited human memory. Viewing human memory as something that is profoundly lacking in its prime function (remembering full and exact registrations of events that happened in the past), they tend to focus on products of memory that presumably fix that fallibility. Yet in doing so, they fail to acknowledge a far more important function of digital media in the act of human memory. If we consider media technologies to be tools for selecting, framing, and encapsulating autobiographical memories—rather than mechanical devices for recording and storing documents or files—they play a constitutive role in the continuous (re)construction of our selves. Technologies of self involve a constant reworking of our relationship with the past; events and reflections are encrypted by technological and discursive devices that actively locate our experience in time and space. Human remembrance, for most software designers, appears to be a natural cognitive process standing ontologically apart from the machine, whereas it is abundantly clear that the brain and the machine are mutually constitutive forces. N. Katherine Hayles has typified the presumed division between the human and the technological as an origin story; she urges a critical understanding of the interactions between the materiality of inscription technologies and the contents they produce.[25] Because this is precisely the aspect missing in the design of contemporary visionary and technical projects, the next section elucidates this interaction.

Memory Storage in the Age of Networked Multimedia

Following in Bush's footsteps, the four contemporary projects of memory machines appear to be predicated on three recurring myths. First, there is the myth that autobiographical memories are indelibly and unerringly stored in the warehouse of the mind, from which they can be retrieved in pristine condition. The second myth is that we have an innate desire to record every single second and facet of our experience in our memory for later retrieval. And third, these projects seem to advance the illusion that memories, recorded and stored in digital databases, can and should be kept separate from the rest of the connected wired world. With the notable exception of Lifestreams, these projects consider individual memories to be private affairs, untouched by the networked collective reservoir of (public) documents and multimedial records.

The question of how digital technologies are affecting the nature of our remembering processes deserves more philosophical contemplation, ethnographic research, and psychological theorizing than the limited context of this chapter allows. Based on the previous analysis of the four projects, it is expedient to turn attention to a few selected aspects of digital memory machines that may help deconstruct the fallacious myths underpinning the engineers' project designs: the computer's morphing abilities, its multimedial and multimodal qualities, and the networked imperative to connect personal to public databases. Unfortunately, engineers and scientists have failed to incorporate new insights in remembering as a cognitive and cultural process into their technology-driven designs. Rather than being self-confined machines for automated recording, storage, and retrieval, multimedia computers can be seen both as technologies of self with surprising creative and affective potential and as technologies of truth (Foucault's terms) pertaining to the worlds of collectivity and social life. The material inscription of signifiers in bits, the convergence of singular media in multimedia machines, and the embedding of personal collections in global networks confront users with profound changes in their cognitive functions and habitual cultural practices. Hence, the question arises: How may the processes of digitization, multimediatization, and googlization impact the construction of memory?

As explained in Chapter 2 and elaborated in subsequent chapters, personal memories are not infallible recordings of past experiences, but they are reconstructions of the past that are filtered, interpreted, and expounded upon through projections and desires. Our ability to create stories of the past does not depend on our ability to recollect precise facts; on the contrary, stories create memories. The morphing nature of episodic memory may lead machines to enhance that quality. According to the reconsolidation theory, memory is a creative amalgamation of fact and fantasy, and the biggest advantage of computers is that they may support this essential feature of episodic memory better than any previous technology. In addition to storing past recordings, we can use digital machines to transform records into new stories that better suit our present understanding of memory and reality. Indeed, although most projects focus on memories as narratives rather than facts, they still tend to stress the accuracy of retrieved recordings. Terms like "accurate," "manipulated," and "false" memories—in opposition to "authentic" and "true" recollections—govern the discourses of psychologists and cognitive researchers, and they also prevail in the design strategy of computer engineers. The obdurate modernist belief in authentic versus manipulated memories obviously echoes in MyLifeBits and other projects.

Instead of confirming fallacious binary frameworks, I have tried to outline how digital tools may help reconceptualize memory as a process etched in time—a process continuously prone to the vagaries of reinterpretation and reordering. If we consider the latest research results coming from neurobiologists and cognitive psychologists, we have to accept human memory as an amalgamation of creative projection, factual retrieval, and narrative recollection of past events. By the same token, we may look upon media tools as creative reminiscing instruments in addition to them being mnemonic aids. Defining memory concurrently as a product and a process, we may acknowledge the versatility and morphing quality of digital memory machines as a positive element that is integral to human reminiscing. Why not create multiple story lines out of stored documents and images while still giving recourse to the original records? Why disregard the creative and transformative potential of memory, as it is such integral element of our identity as human beings? As earlier chapters show, lifelogs may help people keep track of their changing personalities, manipulating digital photographs supports a person's identity formation, mix-and-burn software may customize existing sounds to particular moods, and digitized home movies can offer a reframing of a family's contrived past.

Indeed, digitization of our entire collection of personal documents may elicit new cultural forms and habits besides putting photos in a shoebox or trusting documents to file management systems. Storing, collecting, and retrieving documents—whether pictures, texts, or audio files—are combined mental-technical-cultural processes; a new materiality is bound to affect the mental activity involved in our understanding of self and others, as well as the cultural practices involved in creating and handling memory products. As a result, reminiscing may be defined as a lifelong creative project in which originality is as valuable as authenticity, and in which factual recall naturally supplements imaginative reconstruction. Memory narratives are also fictional stories, just as fiction is always firmly cemented in the crypts of memory. It is the very thing that we love and look for in the digital—its universal fungibility—that will have the biggest impact on what our memories may become, signifying the past as a temporary stage in the enduring process of becoming.

Another persistent tenet held by the creators of digital memory machines is the belief that people want to record every single second and facet of their life's experience for later retrieval. There is little empirical evidence to back up this presumed innate desire. It may not be a coincidence that people's individual cultural memories are structured by the logic of singular media types, woven into specific singular practices, such as photography, making home movies, or compiling audiotapes. Most ordinary users exhibit an unarticulated preference for one medium over another, for instance photography over the taping of moving images. The cultural forms and practices inherent to these singular media technologies unconsciously shape the recording of experiences and thus profoundly affect one's later remembrance of things past. We are not always aware of how the choice of one medium over another quintessentially defines the content of our mediated memories, let alone how it impacts the construction of self-identity over a lifetime. The fact is that the availability or coincidental presence of certain media technologies in one's lifetime often determines one's preference for preserving memories in text, audio, or still or moving images or a combination thereof. It is highly unlikely, though, that with the rise of multimedia computers people would suddenly want to save every life event in every possible sensorial dimension simply because digital technologies allow them to do so.

Yet to what extent will current and future multimedia apparatuses engender a transformation in the process of remembering? In Chapter 1, I argued that mediated memories do not usually serve as exact or complete recordings but as evocative frames. People want a specific sensory inscription that triggers particular emotions or sensations. For instance, a song may help invoke a specific mood, whereas a diary is more suitable to inscribe and recall reflective thoughts. The recording medium once dictated the choice for a particular sensory inscription, but multimedia computers and digital recording devices expand the choices now that digital cameras combined with software packages promote their multiple usage as recorders of sound, text, still pictures, and moving images. I doubt whether people suddenly want to use digital apparatuses to exhaustively record sensory experiences, but they may start experimenting with the synesthetic qualities of the new machines. As users of the Living Memory Box made clear, parents still want to save a toddler's crayon drawings in their shoebox, but their digital equipment enables them to document a child's cute hairdo, first words, and first steps in all their respective visual, auditory, and textual dimensions. The multifunctionality of the digital photo camera unwittingly adds moving images and sound to the recording repertoire of eager parents.

Capturing multiple dimensions of experience—even if inadvertently— may shift a person's propensity to privilege a singular sense in the process of remembering and even lead to new cultural practices. Chapter 5 noted how camera phones do not simply alter the preference for still images in the act of reminiscence by adding text but change social codes and cultural forms of teenage communication as well. In other words, digital technologies potentially change the way we choose to frame our pasts in new sensory modes; but more profoundly, those individual preferences inevitably affect the conventions for remembering and communicating. Sound may undergo a revival in the creation of ego-documents now that audio tags and recorded voices are almost as ubiquitous as image files or text labels. Perhaps personal digital assistants will be deployed to tape dinner-table discussions, which will be disseminated afterward as podcasts—the audio equivalent of e-mailed photographs. We can imagine a large variety of new modes of recording and storing memories, exploiting the new variability of multimedial equipment. It is remarkable to observe how computer engineers bet on multimedia computers as instruments of exhaustive recording and retention, while the real innovation of personal memory machines most likely lies in the combination of new sensorial mind frames with innovative cultural uses.

Finally, most contemporary memory machines (except for Lifestreams) understate the importance of connectivity in their designs of storage and retrieval systems. When Bell introduces MyLifeBits, he states that the human mind ought to work *like* the World Wide Web—the most efficient and effective navigation environment emerging from the digital age. He suggests Google, Silicon Valley's celebrated search engine, as the prime conceptual model for running and activating the human mind. Yet on closer inspection, MyLifeBits designers interpret Google as a mere retrieval tool capable of recovering pieces from a vast but static environment, betraying a rather modernist notion of human memory as well as a remarkably limited understanding of the capabilities of search engines. In Bell's conceptual model, memories are fixed entities patiently awaiting their retrieval from the shelves of the mental library, triggered by a coincidental or conscious thought. Today, these assumptions have been discredited in favor of memory as a complicated encoding process, where memories are preserved through and affected by elaborate mental, social, and media schemata. Specific needs, interests, or desires on the part of the rememberer significantly structure the content of memories while at the same time they define communication and expression. Human memory is a flexible agency through which identity development, and thus personal growth, is made possible. Microsoft engineers ignore the inherent transgressive qualities of human memory; they stress the functionality of memory as a storage machine, at the expense of its creative, communicative, and connectionist capabilities. Moreover, Bell's metaphorical equation of memory to an intelligent software agent (that is, Google) does not do justice to the transformative power inscribed in this search engine; Google's power lies not in its ability to search fixed sets of databases, but in its ability to navigate a person through a vast repository of mutant items, yielding different content depending on when and how they are retrieved, reshaping the order of its data upon each usage. Two unique Google qualities—its navigability and its constantly changing inventory—are conspicuously absent in MyLifeBits and other projections of future memory machines.

Ignoring the inherent mutability of human memory may seriously hamper the high ambitions set by Microsoft and other designers. Projects like Shoebox, the Living Memory Box, and MyLifeBits tend to regard software programs as self-contained nostalgia machines—jukeboxes of individual memory—that perform a mnemonic function: retrieve that particular song or recall that image. But thanks to the networked computer, memory becomes more of a topological skill: to navigate personal memory not only highlights the process of remembering but also allows the user to make connections that would have never been discovered without the computer. For instance, a person may detect patterns in his or her musical preference or a specific development in the family's vacation photographs over the years, particularly if relating personal items to historical trends. Chapter 6 explained how various types of home movies and videos, after being juxtaposed and connected to publicly available documents via digital media technology, led to ambiguous interpretations and innovative representations of a family's past records. Topology and navigation, in addition to retrievability, make the memory process a more intriguing effort than ever before; the networked computer is a performative agent in the act of remembering. Reconnection is perfectly compatible with recollection, as individual memories are always inherently embedded in cultural contexts. The googlization of memory, in Microsoft terms, is a callow conceptualization of what networked modalities may produce; the emergence of new genres that connect private memories to reflections of others or to public resources—and thus produce new thoughts—is the true innovative potential of a digital memory machine.

Contemporary designs for personal memory machines assume technological innovations to provide a new model for human memory, such as the computer or the World Wide Web, but they fail to acknowledge the mutual shaping of human memory and machine. A basic flaw in the digital machine's software design is that it models the brain after the computer and assumes the private mind to be searchable like the World Wide Web is searchable; personal memory is thus considered a repository that is completely separate from collective memory. Projects like Shoebox and the Living Memory Box concentrate on the collection of personal memory items only. Lifestreams and MyLifeBits duly acknowledge the interlacing of personal and public records (or records sent by others), but at best they address the issue of ubiquitous connectivity in terms of privacy infringement and personal integrity. However, they fail to recognize the innovative potential of new digital environments; the World Wide Web opens up space for new cultural practices fulfilling a social need for connecting the self to larger contexts of community, society, and history. Memory is neither exclusively cemented in the recall of individual experiences nor in the remembrance of collective experiences, but a human being has a vested interest in connecting both poles if he or she wants to pursue personal growth.

Networked digital technologies have the ability to link up personal memory to public mediated materials, hence eliciting insights in the interconnection between self and the world. For example, the diary or the scrapbook in its analog form serve as a reflexive instrument in the contained universe of a person's life. The new potential of the networked computer is not, as Bell would have it, scanning all words into the computer in order to render one's personal testimony searchable via keywords. The real innovation of the computer is its ability to allow a new type of diary (for instance a lifelog), which networked materiality preconditions the linking of private reflections to public scrutiny, opening up personal reflexivity to invite reciprocation by others. As illustrated in Chapter 3, privacy and publicness are not mutually exclusive concepts in the lifelogs of teenagers, and Alzheimer's patients even strive for publicity and intimacy. In other words, the working of personal cultural memory changes in the face of networked machines, a transformation that calls for a renewed awareness of the relation between personal and collective memory.[26]

Let us return finally to the problem of my friend, cited at the beginning of this chapter, facing the storage and retrieval of innumerable memory-infested digital files. Would he be significantly helped by projects like Shoebox, the Living Memory Box, Lifestreams, and MyLifeBits? Would these software projects help him restore order in the plethora of digital memorabilia and allow him more time to relive captured moments? Confirming my hypothesis, my friend answered that he was not exactly waiting for sophisticated programs or intelligent agents to assist him in finding particular items and turn them into smooth multimedia presentations. "The funny thing is," he said, "I am not very keen on retrieving the experiences I recorded. The value of my personal digital collection is situated first and foremost in the fun of recording and collecting and perhaps second in knowing that these files are somehow stored, in coding, even if I will never retrieve them." His conclusion that he treasures the act of collecting more than his actual collection is not something unique to life recorders in the digital age; rather, it is quite analogous to the woman who keeps her love letters wrapped in ribbon in an attic but never or almost never looks at them. My friend's digital recordings, apparently, had already served their purpose in the act of memory, even though as objects of memory, they may never materialize beyond their coded stage. While building his new digital environment, he had noticed how new tools had helped him to organize his everyday world and to find a new way to inscribe his personal life in dimensions of time and order.

The performative nature of memory is, I believe, much underexposed in current research on memory machines. Memories are narratives as well as artifacts, performances as well as objects—things that work in every day lives and cultures of people. In their search for the perfect digital memory machine, engineers and entrepreneurs have systematically focused on the products of memory and ignored the role of technologies in the active staging of memory. They assume mind and machine to be separate entities, one serving as the model for the other, generally disregarding their perpetual mutual encroachment. Computer engineers, as designers of digital tools, could profit from neuroscientific research to obtain insights in how autobiographical memory works and evolves; they could also benefit from the wealth of ethnographic inquiries into the individual's use of both material and digital memories in the process of collecting, storing, and retrieving.[27] Working at the crossroads of various disciplinary perspectives, we may try to find out how digital materiality impinges on our everyday habits of preserving and presenting personal heritages. Digital technologies seem to promote a different materiality that both complements and partially replaces analog objects embodying memory; most important, they shape the very nature of remembering as they become (literally) incorporated in our daily routines of self-formation. In spite of current project designers' projections, the ultimate goal of memory is not to end up as a Powerpoint presentation on your grandchild's desktop; the ultimate goal of memory is to make sense of one's life.

Epilogue

I must have been around twenty years old, when I saw Francis Picabia's painting *The Acrobat* in a Scandinavian museum—I cannot remember whether it was Oslo or Stockholm. What struck me in the painting were the thick brush strokes accentuating the unnatural upside-down posture of the acrobat. The poster I bought of this picture unfortunately attenuated the brush strokes, but it nevertheless covered the wall of my student room for some time. More than ten years later, while browsing in a bookstore in Los Angeles, I hit upon a reproduction of Picabia's painting in an art book. I only remember that particular browsing experience—as distinct from hundreds of hours of browsing in American bookstores—because of the reproduction. Not until recently, almost twenty-five years after first seeing the painting, did these two moments come back to me via an old photograph of myself on a couch in my student room, in which the poster on the wall figures as background. The friend who had taken the picture had kept it in her shoebox for many years; when she found the photograph, she scanned it and e-mailed it to me. I looked on the Internet for a clean reproduction of my beloved acrobat and reworked both images into a humorous story about my peculiar run-ins with this painting, which I then e-mailed back to my friend.

Mediated memories, as proposed in the first chapter, are the activities and objects we produce and appropriate by means of media technologies for creating and re-creating a sense of our past, present, and future selves in relation to others. Photographs, posters, and digital reproductions of the Picabia painting all played a role in my memory act, involving a variety of material mediations and triggering a number of mental activities and affects. Rather than being mere relics, they serve as input in a continuous memory discourse with variable output at different moments. In Chapters 1 and 2 of this book, I designed a model to analyze the interlocking of media and memory as moving along two major axes. The horizontal axis—the axis of relational identity—places the manifestation of cultural memory at the crossroads of self and others, individual and collective, private and public. The vertical axis laid out the dimension of time: the integration of past and future in the present, the mixture of recollection and projection, and the fusion of preservation and creation. In Chapters 3 through 7, this dual model was applied to concrete mediated memories at a time of technological and cultural transformation. Here I sketch some tentative conclusions with regard to the model.

Evaluating the Model

Mediated memories are reciprocal in nature; they mediate between self and others. Media technologies involved in creating personal memories are not simply used to build up a personal reservoir of memories, but their function is concurrently formative, directive, and communicative. They enable the self to grow and mature, to give meaning and direction to one's past and present, while at the same time they allow a person to communicate with others and test common grounds. Whereas the Picabia poster and art book reproduction once reified my experience of viewing the painting, the photograph of the poster became the catalyst of a memory experience. The photograph served as a mnemonic device helping to imagine my former self; it also had a directive function in the formation of identity; and it engendered a communicative meaning as exemplified by the online image exchange. Digital environments, by nature of their networked condition, amplify and encourage connections between self and others (and culture at large). Chapter 3 showed how lifelogs offer a new hybrid form that combines formative, mnemonic, and directive functions, particularly in supporting teenagers' and Alzheimer's patients' volatile sense of self.

Mediated memories also form the linchpin between individual and collective culture. Acts and objects of memory, as argued in Chapter 1, have always been a playground for sorting and appropriating cultural items. Digital technologies empower consumers of culture, arming them with advanced means to construct collective identities. Taking photographs and buying posters are both acts of appropriation: by defining their idiosyncratic choices in the sea of cultural products, individuals mark their preferences but also communicate a sense of belonging to a (peer) group or a (sub)cultural league. Chapter 4 described how digital music files lend themselves to easy exchange of pop songs and thus contribute to the formation of one's personal taste in relation to collective musical heritage. In more than one way, the distinction between individual and collective memory evaporates as individuals obtain and wield the (digital) instruments that lie at the heart of cultural production. Home movies inspire television series as much as feature films inspire home video productions (Chapter 6). In short, personal and collective cultural products are almost seamlessly interwoven into the symbolic fabric of everyday life.

Memories also mediate between what is private and what is public, concurrently modifying the meanings of privacy and publicness. The development of every new medium—beginning with print—has reconstituted the boundaries between public and private life. Dissemination of personal memory is increasingly an online activity: from the creation of lifelogs to the distribution of virtual images, the already thin line between private and public has only become more diffuse. Personal cultural memory, of course, was never squarely located within the private realm, because individuals made conscious decisions to publish their shoebox contents. That private shoebox is gradually integrated in a global, digital bazaar of documents, music, and pictures, where files appear almost indistinguishable. So when I put my picture of the Picabia poster on the Internet, cutting and pasting freely from digitized museum reproductions, I should not be surprised by claims of copyright infringement. The opposite is also true: a specific entry in your lifelog or a picture sent by a camera phone (Chapter 5) may later show up in public contexts as a result of having shared it via a public forum, even if unwittingly. Personal and collective cultural items thus become progressively interwoven in memory discourses.

The vertical axis in the proposed model signifies how memories mediate between past and future. One of autobiographical memory's most important functions is that of self-continuity: in spite of the body being a constant physical transmutation of cellular tissue, appearance, and

thoughts, we always look for continuities in our identity—to reassure ourselves that the trunk of our present life tree has historical roots and future
branches. And yet, we can only gain access to the past through the present;
my memory of the Picabia image is an act in the present, even though the
objects that trigger my memory were made a long time ago. As elucidated
in Chapter 6, memory objects such as home movies move up and down
the time line. People may create movies as future memories, and they edit
moving images shot in the past to attune them to current views of family.
Our memories of the past change along with our expectations of the future; memories are constantly prone to revision just as memory objects are
constantly amenable to alteration.

Recollection and projection are essential ingredients of our present
construction of memory and identity. Mental images of who we are result
from a combination of recall and desire, which are in turn incentives to remodel our past and fashion our futures. Even if I chose the Picabia poster
because I simply wanted to decorate my sparsely furnished student room,
the story written twenty-five years later presented my choice as a sign of
my emerging preference for European avant-garde painting. Partly recall,
partly projection, I construct an image of my former self in the present:
I want to remember myself as an art lover because it fits my current self-
image. Chapter 5 showed how personal photographs and memory narratives
could be easily manipulated, particularly in the digital age, as photoshop
software allows for untraceable modifications in our (physical) appearance.
The digital collections on our desktops are brimming over with constant
reminders of our former selves, but by the same means they provide the
tools to help us shape our idealized image—a projection of who we want
to be and how we want to be remembered.

Mediated memories in the digital age are creative reconstructions as
much as documentary scenarios of what happened. Memory has become
an interesting amalgamation of preservation and creation: that which remains stored in shoeboxes or other containers serves as input for the invention of new memories. Interlacing my memory narrative with digital
images of an old photograph and some Picabia reproductions enhanced
my pleasurable experience of mixing preservation and imagination. In
constructing my memory of the Picabia painting, I am not exactly interested in which aspects of my narrative are accurate and which are false. Pictures, diaries, home movies, and recorded music were never plain evidence

of our former states of being or past experiences anyway. Memory objects created in the past provide input for current interpretations of present and future life, a status to which digital media objects lend themselves better than their analog precursors. Memories are creative material, as they lead to therapeutic and productive new insights into our selves and may result in innovative cultural forms, such as DVD recordings of a family dispute. Software designers are very slow to pick up on the creative side of autobiographical reminiscence, as they keep stressing the importance of memory preservation and storage in preformatted scrapbooks.[1]

An Interdisciplinary Approach to Memory

The model proposed and advocated in this book, a model structured along the two axes of identity and time, strongly promotes a multidisciplinary approach to memory. Mediated memories means our memories are embodied by individual brains and minds, enabled by the technologies and material objects that render them manifest, and embedded in social practices and cultural forms. But why do we need such a complex model? What are the advantages of combining diverse and seemingly irreconcilable definitions of memory? Why would academics from various backgrounds want to engage with such a broad and multilayered model before they have answered all or even most of the pressing questions emanating from their disciplinary perspectives? Why would it be advantageous for neurobiologists or cognitive psychologists to turn to cultural theorists or social constructivists (or vice versa) to deepen their insights in how memory works? My answer to these questions is simple: academics cannot afford to ignore each other's research paradigms in developing new horizons on memory in the digital age. Because we are facing new challenges posed by complex technical-cognitive-cultural phenomena, we can hardly rely on old models and metaphors to advance our thinking. The mutual shaping of mind, technology, and culture is an important premise of the proposed model; its modest goal is to pull together relevant findings from these various perspectives to see how they can help redefine immanent multifaceted problems in memory research, such as the problems of morphibility, multimediatization, and networked connectivity.

The reconsolidation theory, developed by neuroscientists and explained in Chapter 2, takes as its premise the mutant nature of autobiographical memory; personal memory is tweaked and turned whenever an object cre-ated in the past is recalled through the mindset and conditions of the pres-ent. Forgetting and remembering are functions of the mind's inclination to produce creative recollections mingled with projections. Cognitive psychol-ogists, for their part, have convincingly proven how memory not only morphs but can also be morphed. As argued in Chapter 5, exposure to ma-nipulated misinformation or doctored photographs can change an individ-ual's recollection in powerful ways. But memory morphing can hardly be studied apart from the phenomenon of media morphing—a term that refers to digital multimedia's capacity to smoothly convert one coded representation into another. Media morphing may pertain to the software accommodating the invisible makeovers of photographs in digital video footage; it may also apply to the technical ability to make old footage look pristine or blend personal and public images. The morphibility of human memory is not simply supported by media technologies, but also, to a large extent, it is enabled by and inscribed in digital tools. My Picabia memo-ries, of course, emanated from the brain-mind, where I had conjured up specific (even if changing) mental images of *The Acrobat.* Due to my "dig-ital memory machine," I could effortlessly merge former images with con-temporary reproductions from posters, photographs, and web images, thus creating an interesting amalgam of recall and projection.

New digital tools thus engender new types of memory, or, mutatis mutandis, cognitive mechanisms give rise to new technological manifesta-tions of memory. For instance, computers and software promote the multi-modality of memory: the combination of different sensory modalities such as visual, auditory, and haptic impressions. Multimediality refers to the si-multaneous use of multiple media formats, such as text, graphics, anima-tion, pictures, videos, and sound, to present information. Research into the multimodality of memory—an area largely contingent upon the use of multimedia computers—is becoming a fruitful subject for cognitive psy-chologists and neuroscientists involved in memory research. For the design of human-computer interfaces, engineers rely heavily on cognitive research into multimodal perception. Few, if any, insights in this area of multi-modality have yet been applied to the study of personal memory. As dis-cussed in Chapter 7, computer scientists working on new memory machines still generally depart from the assumption that we need these machines to store accurate complete memories, and they have only barely started to discover the creative potential of hardware and software into their new machines. It will be interesting to experiment with the multimodal potential of the new technologies in terms of memory enhancement.

Another major omission in the design of these digital memory machines is their remarkable indifference to contextual use; personal memory is always embedded in sociocultural forms and practices, which are partly responsible for its shaping. When we store a picture for later retrieval, this act is part of a social custom to save and store photographs for later reminiscence; the photo album or shoebox is the conventional cultural form in which we preserve memories. The photograph of my student room featuring the poster was preserved in somebody else's shoebox, but was recollected because of a new connective custom: to electronically distribute pictures to friends—a social practice grounded in the networked condition of connectivity. It may be safe to assume that the current transformation of technology in conjunction with sociocultural forms and practices poses a challenge to the human mind and how it selects, understands, and remembers information about the self and relates this information to others and to culture at large. A camera phone, as elucidated in Chapter 5, literally merges the functions of memory and communication; the resulting image-text is neither an exclusive memory object nor an ephemeral byproduct of a casual conversation. New social practices and cultural forms emerge at the crossroads of memory and media, and even if we cannot yet label them, it is important to mark their appearance. It is too early to tell how cultural forms will develop further into the digital age, but there is no denying that the functions of memory, communication, and identity formation are becoming evermore intertwined.

Recognizing the connection between memories' embodiment, their enabling technologies, and their embedding in social practices and cultural forms motivates my call for an interdisciplinary approach to memory. Rather than considering technology the engine of a multifaceted transformation, this book opted for a more complex analysis of personal cultural memory, one that advocates neither the teleology of change nor a compartmentalized patchwork of separate evolutions. Instead, I show how results from neurobiological and cognitive research may complement social-constructivist paradigms concerning media technologies and how software designers may profit from insights in sociocultural patterns of media use. By defining memory as a phenomenon that needs biological-cognitive, material-technological, and sociocultural scrutiny, I do not mean to discount research paradigms pursued in each of these respective academic disciplines. On the contrary, my prime intention is to transfer insights in one area onto the next, to bring about connections that would otherwise remain unnoticed, and perhaps inspire new and innovative conceptualizations of personal memory. Cultural theorists thus far largely ignore the great advancements in neurobiological theory and cognitive psychology; neurobiologists show little interest in the meaning of material-technological objects as agents of memory; and social constructivists emphasize technology but have a lukewarm appetite for including cultural dimensions of change.

This book has no pretension whatsoever to provide a manual for collaborative research projects involving neuroscientists, sociologists of technology, and media theorists. Its inventory of shoebox objects, each representing a prism onto the process of cultural and technological transformation, is an exercise in expansive thinking, theorizing the object of memory across conventional disciplinary boundaries. As we are entering the digital age, collaboration and cross-fertilization may become increasingly more urgent, because we are faced with phenomena that concurrently involve various levels of memory research. Phenomena such as morphibility, multimediatization, and connectivity (or googlization, see Chapter 7) can no longer be understood as either mental or technological or cultural manifestations of memory. A profound understanding of these phenomena requires the help of multifocal lenses instead of monocles.

Memory Studies in the Future

One of the challenges of scientists is to open up well-defined problems to ever more complex questions. The more narrowly neurobiologists and cognitive scientists identify the question of how human memory works on the genetic, molecular, and organic levels, the more challenging it is to expand the focus to memory's interaction with machines, objects, and sociocultural schemes. But how can we reconceptualize the problem of memory in a way that accounts for evolving transformations in the areas of (neuro)cognition, digital technology, and culture? The acknowledgment of morphibility, multimediatization, and connectivity as emerging and pressing themes in memory studies may have considerable consequences for its epistemology and ontology, warranting further interdisciplinary efforts.

The morphibility and multimediatization of memory and media are not distinct phenomena, as they are interwoven at every imaginable level. Much of the current research on memory is grounded in parameters of authenticity versus artificiality, of truthful recollection versus manipulated or false remembrance, of comprehensive versus selective memory. If we accept the morphing and multimodal nature of memory, defined as a bio-cognitive process entwined with its technological-material manifestations and sociocultural practices, then we can no longer build upon these bipolar frameworks. Every memory is mediated by a self, with the help of artificial instruments. Truthful memories are naturally morphed; even the most vivid, detailed, and documented memory is necessarily a selection of modalities and thus never comprehensive. To such a concept of memory, terms such as "true" and "false" no longer apply (if they ever did), because memory is intrinsically a mediated phenomenon—mediated by the vagaries of mind, technology, and culture. Rather than judging memory's truthful content by asking what happened, a necessary corollary to that question is to assess its specific manifestation and (in)consistency as it evolves over time and as it is exposed to alternative versions of true memories. Perhaps we should slightly change our focus to questions such as: How do we know in the age of digital media which information is coming from whom and with what intent? How do we recognize and judge information that presents itself as personal memory or historical fact? For instance, cognitive psychologist Elizabeth Loftus cautiously introduces such expansive definition when claiming that memory is not the sum of what people have done but also encompasses "what they have thought, what they have been told, what they believe."[2] I would add to this definition that personal cultural memory is also what people have selected to record and perceive; what they have documented and how they chose to document it; what stories, sounds, and images they were exposed to and how they have re-created them; how they want to remember and be remembered; and how and in what form they have presented and communicated these acts of remembrance to others.

The unlimited potential of digital technologies to combine and fuse sensory modalities may also redefine our research questions. I already noted the ongoing cooperation between cognitive scientists and computer engineers in the area of human-computer interfaces, but cultural theory plays virtually no role in this joined effort. Some of the most promising products resulting from this collaboration are the designs of new virtual environments that serve as simulations of real experiences.[3] Multimodal 3-D environments are currently tested in the field of psychotherapy, for instance to treat phobic disorders, or in courtroom settings to reconstruct actual crime scenes.[4] In both cases, digital apparatuses help human memory to perform. The vulnerability of autobiographical memory to guided imagination in addition to the capabilities of new digital media to paint memories in digital multimedial productions may necessitate new standards for litigation or successful therapy. Will we accept a multimedia reconstruction in the courtroom, for instance a graphic 3-D reconstruction of a crime scene or the scene of an accident, knowing that a multimedial representation also imposes and induces a certain version of what happened? How is human recollection affected by multimedial input? How do cultural forms, such as televised crime-scene investigations, influence concrete personal memories of criminal scenes? How do computer games featuring virtual chases and enemy shoot-outs affect (or interfere with) the treatment of post-traumatic stress disorders?

If morphibility and connectivity are becoming the default mode of mediated memories in the digital age, we need to adjust our research questions accordingly. The ontology of memory, like the ontology of media, used to be firmly rooted in metaphors such as the archive or the repository. Memory was considered something we have or lack, retain or lose, but it was never a state of becoming. In conjunction with this metaphor, we speak of retrieving or recalling past events, implying that the thing to be recalled is already there waiting to be retrieved. But with the implementation of digital media tools in the everyday construction of memory, our very concept of how memory functions is technically and metaphorically grounded in different parameters. Does this mean we should now define autobiographical memory by its "track changes" mode in addition to its "save file" mode? Acknowledging the track changes mode as a valid mindset may boost the appreciation of new cultural practices, such as the lifelog and the photoblog, which favor the dynamics of recording over the desire to freeze the past. Adjunct to its morphibility, the networked lifelog emphasizes memory's connective function in addition to its commemorative use. A collection of digitized personal files stands in perpetual dialogue with the infinite basin of files on the Internet—a repository that is by definition dynamic. Instead of retrieving memories, we may now speak of navigating memory files, concurrently changing the emphasis from static sources to mutant items and from a closed mental space to an endless, open reservoir of (private and public) dossiers. Indeed, accepting the mutability and connectivity of memory items entails we recognize the ability of the mind and our navigational instruments to change and be changed, and this in turn affects the content and status of memories. Memories are never a simple inheritance from the past: people make media to shape memories, and memories shape people to make media. Recollection is also a form of reconnection, as mental and cultural processes are involved in the digital restructuring of personal memory.

Facing contemporary phenomena such as morphibility, multimodality, and connectivity augments the urgency to rethink research questions and conceptual models to study human memory. Through the prism of mediated memories, I offer an analytical tool that supports cross-disciplinary thinking. So far, academic approaches to memory have focused either on autobiographical or on collective memory; academics have examined memory either as a mental activity or as a cultural activity, generally regarding media as instrumental aids to support these activities. The proposed model tries to bridge these schisms by defining memory as an intrinsically mediating activity: it acts as a go-between between individual and collective identity, between mental and cultural imaging processes, and between the past and future. It is remarkable how some of the central terms used in memory research, terms such as "identity," "trauma," "manipulation," "heritage," "experience," and "images," are appropriated very differently in divergent disciplines. And yet their interrelatedness is crucial to neurobiologists, psychologists, media theorists, and historians alike, and it becomes even more so as their changing meanings epitomize larger historical transformations.

The morphibility or changeability of identity as a feature not only of our minds but also of our bodies can hardly be confined to its mental or technical strata only. The volatility of cultural identity—as a result of migrant flows and increased mobility—is immanent to this same condition, as people's psychological and physical view idea of self can hardly be separated from their idea of belonging to a group (ethnic, generational, and so forth). Most ardently illustrated by the artwork of Nancy Burson's race machine, introduced in Chapter 5, which promotes the morphing of racial features in bodies, pictures, and cultural identities all at the same time, I have argued how the morphing of memory is part of a more general cultural condition—a tendency to control and stimulate the active shaping of identity. People identify themselves and the way they look by studying the looks of ancestors, relatives, and peers, creating continuity by discovering roots. Media technologies are a means to help us discover and create, reconstruct and re-create a sense of self in the context of historical continuities.

Trauma is another term monopolized by psychiatrists and psychologists but inevitably appropriated by historians and cultural theorists to theorize the propensities of human memory. Psychologists, for instance, are interested in how the experience of war or violence in a child's past may affect that person's present and future state of mind, whereas historians transfer the term trauma to study collective wounds caused by war or violence and how we cope with them in revising our histories. The proposed connectionist model encourages the study of trauma as a mediated experience: the Afghan refugee's angst or Somalian war child's trauma are both mediated by the (lack of) personal photographs showing a happy family and by the news photographs documenting scenes of continuous warmongering. The exposure to—but also the manipulation of—public photographs with the help of media technologies creates building blocks in the construction of personal identity and cultural heritage. By the same token, the (manipulability of) pictures and personal narratives may be a creative tool in the process of healing and coming to terms with traumatic memories. "Connectivity" and "communicating" may be new key words in the definition of personal cultural memory, terms that remind us of the relevance of memory to our well-being as humans.

When researchers face new phenomena and thus new challenges, they tend to look for common ground that may help them test proven hypotheses against new paradigms. In the future, memory research may include horizons that currently escape a scientist's immediate disciplinary scope. That is why I never attempted to define what memory is but instead focused on its mediatedness. In designing an expansive model for memory research, I worked from the assumption that memory always serves a purpose and that it works toward a certain goal: we remember because we want to make meaning out of life. Memory makes meaning by mediating between disparate abstract and concrete entities: the self and the world, the mortal individual and the immortal collective, the family's past and the future generation. Memory is only purposeful if it mediates and thus connects unknown entities to known ones; we trace memory as a dynamic mechanism, moving between ends and manifesting in changing forms. That is why I chose to examine mediated memories at a time of technological and cultural transition: not because digital technology is in any way more conducive to memory than analog technology, but because a state of transformation is better able than a steady state to show how mental, technical, and cultural aspects of memory alter in conjunction. In the future of memory research, it will be a challenge to attune any model to mutual interests and insights of collaborating researchers.

Memory is no longer what we remember it to be, but then, memory probably never quite was how we remembered it and may never be what it is now. The present is the only prism we have to look through to assess memory's past and future, and it is important we look through this contemporary prism from all possible angles to appreciate memory's complexity and beauty. As I have tried to show in the preceding chapters, neurobiological and cognitive psychological theories of autobiographical memory are surprisingly akin to social-constructivist and humanities approaches to cultural memory, and yet there is little incentive for collaboration until new phenomena call for new tactics. *Mediated Memories in the Digital Age* lays the groundwork for an interdisciplinary approach, with the hope that its framework will be built upon and expanded.

Notes

1. Although there are some subtle differences between the words "personal" and "autobiographical" in connection to memory, most psychologists and neuroscientists use them interchangeably. For stylistic reasons I alternate between the terms "autobiographical" and "personal," yet I strongly prefer the term "personal" in connection to cultural memory, for reasons to be explained later in this chapter, toward the end of the second section.

2. For more information on the development of the autobiographical self (in relation to the biological self) see Antonio Damasio, *The Feeling of What Happens: Body and Emotion in the Making of Consciousness* (Orlando, FL: Harcourt, 1999), 222–33. I further elaborate on this theory in Chapters 2 and 4.

3. Susan Bluck, "Autobiographical Memory: Exploring Its Functions in Everyday Life," *Memory* 11, no. 2 (2003): 113–23.

4. Katherine Nelson, "Narrative and Self, Myth and Memory: Emergence of the Cultural Self," in *Autobiographical Memory and the Construction of a Narrative Self: Developmental and Cultural Perspectives*, ed. Robyn Fivush and Catherine Haden (Mahwah, NJ: Lawrence Erlbaum, 2003), 3–25. In this article, Nelson lists the different levels of self-understanding psychologists have long recognized: in addition to a cognitive and social sense of self, experts have distinguished representational and narrative levels of understanding. See also Katherine Nelson, "Self and Social Functions: Individual Autobiographical Memory and Collective Narrative," *Memory* 11, no. 2 (2003): 125–36.

5. Nelson, "Narrative and Self," 7.

6. Nelson, "Self and Social Functions," 127.

7. See Qi Wang and Jens Brockmeier, "Autobiographical Remembering as Cultural Practice: Understanding the Interplay between Memory, Self and Culture," *Culture and Psychology* 8, no. 1 (2002): 45–64, 45.

8. Cultural differences of *remembering* between parents and children of Chinese and American descent have been studied by Wang and Brockmeier, "Autobiographical Remembering as Cultural Practice," 54–60. Differences between American and

Japanese ways of memory recording—in this case amateur photography—is the subject of Richard Chalfen and Mai Murui, "Print Club Photography in Japan: Framing Social Relationships," *Visual Sociology* 16, no. 1 (2001): 55–73.

9. My definition comes close to what Annette Kuhn terms "memory work" in "A Journey through Memory," in *Memory and Methodology,* ed. Susannah Radstone (Oxford: Berg, 2000), 183–96. She defines "memory work" as an "active practice of remembering which takes an inquiring attitude towards the past and the activity of its (re)construction through memory" (186). Memory work is in fact a series of activities (inscribing or recording, interpreting, narrating, recalling, and so on) that may involve a number of memory products (ranging from strings of hair or a child's tooth to pictures, videos, and diaries). For reasons explained later in this chapter, I restrict my focus (unlike Kuhn) to memory objects resulting from the interference of media tools.

10. Analyzing the nature of (digital) pictures on our mindset and cultural activity, Ron Burnett in *How Images Think* (Cambridge, MA: MIT Press, 2005) observes that "the intersections of creativity, viewing, and critical reflection are fundamental to the very act of engaging with images in all forms' defying the myth of the passive viewer" (13).

11. Think, for instance, of popular cultural products like the *Bridget Jones's Diary* and *America's Funniest Home Videos*, which both expound on personal memory forms, even if some of these forms are fictional.

12. Maurice Halbwachs, *On Collective Memory* (Chicago: University of Chicago Press 1992), 34. Halbwachs's famous work was originally written in 1925 as *Les cadres sociaux de la memoire* and published in 1950 as *La memoire collective* by Presses Universitaires de France, Paris.

13. The reprint of *La memoire collective,* in 1968, also contained Halbwachs's unfinished essay "La memoire collective chez les musiciens."

14. See, for instance, Paul Connerton, *How Societies Remember* (Cambridge: Cambridge University Press, 1989).

15. Such connections can also be built around archival systems; these so-called communities of records signify how collectivity is construed by means of (re)positioning documents in archives. See Eric Ketelaar, "Sharing: Collected Memories in Communities of Records," *Archives and Manuscripts* 23 (2005): 44–61.

16. David Gross, *Lost Time: On Remembering and Forgetting in Late Modern Culture* (Amherst: University of Massachusetts Press, 2000).

17. According to Gross, in *Lost Time,* the individual is still free to remember outside these general frames and create a divergent or oppositional memory that would "recall things about the past that are not commonly thought about and perhaps not even missed in the population at large" (134).

18. The observation that individual narratives, or stories of personal experiences, have become increasingly important in twentieth-century Euro-American

culture is corroborated by quite a few psychologists. For instance, Katherine Nelson, in "Self and Social Functions," remarks that in current culture "the vanishing of common communal narratives is replaced with a cacophony of personal stories, which makes it necessary for individuals to each add their own unique self story" (134).

19. Andrew Hoskins, "Signs of the Holocaust: Exhibiting Memory in a Mediated Age," *Media, Culture and Society* 25, no. 1 (2003): 7–22, 10.

20. Steven Spielberg established the Survivors of the Shoah Visual History Foundation right after completing his film *Schindler's List*. The mega-project initiated and supervised the audio-visual recording of over fifty thousand testimonies of Holocaust survivors from fifty-seven countries and in thirty-two languages. More information on the project can be found at http://www.vhf.org (accessed December 21, 2006).

21. Andreas Huyssen, *Twilight Memories: Marking Time in a Culture of Amnesia* (New York: Routledge, 1995), 255.

22. Assuming its representational nature, Huyssen firmly locates memory in the realm of culture rather than in the realm of cognition or sociality. He ipso facto rejects a potential distinction between living and artificial memory; if memory is by nature a representation of the past, it can only be examined by means of its discursive or material manifestations. Considered as representations of the past, the distinction between individual and collective memory becomes less distinct. Although I am largely sympathetic to the idea of memory as inherently a representation of our individual and collective pasts, I also find this concept lacking. Collective memory is often caught in terms of content or message, whereas I prefer to regard it as a stilled confrontation between individuality and collectivity in which intention and control are keywords.

23. See Ernst van Alphen, "Symptoms of Discursivity: Experience, Memory, and Trauma," in *Acts of Memory: Cultural Recall in the Present*, ed. Mieke Bal, Jonathan Crew, and Leo Spitzer (Hanover, NH: University Press of New England, 1999), 24–38.

24. Jan Assmann and John Czaplicka, "Collective Memory and Cultural Identity," *New German Critique* 65 (1995): 125–133, 126.

25. Aleida Assmann, "Four Formats of Memory—From Individual to Collective Forms of Constructing the Past" (lecture, presented at the conference "Theatres of Memory," University of Amsterdam, January 28, 2004).

26. I should note here that Aleida Assmann pairs off individual with social memory on one end and groups political with cultural memory on the other; as a historian, she is mostly interested in the latter. The term "social memory" is more or less appropriated by sociologists, who have delineated the field of social memory studies as their turf, even if recent theories insist on the close interrelationships between various types of memory research. For an illuminating overview, see Jeffrey K. Olick and Joyce Robbins, "Social Memory Studies: From Collective

Memory to the Historical Sociology of Mnemonic Practices," *Annual Review of Sociology* 24 (1998): 105–40. Olick and Robbins define social memory studies as "a general rubric for inquiry into the varieties of forms through which we are shaped by the past, conscious and unconscious, public and private, material and communicative, consensual and challenged. We refer to distinct sets of mnemonic practices in various social rites, rather than to collective memory as a thing" (112).

27. On the idea of the Anne Frank museum as an act or "performance" of collective memory, see Sonja Neef, "Authentic Events: The Diaries of Anne Frank and the Alleged Diaries of Adolf Hitler," in *Sign Here! Handwriting in the Age of Technological Reproduction,* ed. Sonja Neef, José van Dijck, and Eric Ketelaar (Amsterdam: Amsterdam University Press, 2006): 23–50.

28. Instead of speaking about "retention" or "loss," Andrew Hoskins, in "New memory: Mediating History," *Historical Journal of Film, Radio and Television* 21, no. 4 (2001): 333–46, concludes that it is more appropriate to consider our relation to the past "in terms of its mediation or remediation in the global present" (335).

29. Francis Yates, in *The Art of Memory* (Chicago: Chicago University Press, 1966), points at the significance of photography in the devaluation of memory. The advent of mechanically reproduced images was thought to lead to the destruction of truth and was therefore thought to undermine human memory.

30. Media theorist Walter Ong, in his renowned book *Orality and Literacy: The Technologizing of the Word* (London: Routledge, 1982), argues that a memory stilled in words, whether spoken or written, is just as "technologized" as a memory packaged in electronic images; both combine external technologies and internal techniques to help structure our remembrance.

31. See, for instance, Mitchell Stephens, *The Rise of the Image, the Fall of the Word* (New York: Oxford University Press, 1998). Stephens argues that the popularity of the image could only rise at the expense of the written word. Elsewhere, I have extensively countered Stephens's and others' assumption that writing and imaging should be defined as opposites in a communicative system. See José van Dijck, "No Images without Words" in *The Image Society. Essays on Visual Culture,* ed. Frits Gierstberg and Warna Oosterbaan (Rotterdam: NAi Publishers, 2002), 34–43.

32. For a basic introduction to McLuhan's ideas on memories as extension of the body, see *Understanding Media* (New York: McGraw-Hill, 1964), particularly chapters 1, 2, 28, 30, and 31.

33. Pierre Nora, "Between Memory and History: Les lieux de memoire,"*Representations* 26 (1989): 69–85.

34. Jacques Le Goff, *History and Memory* (New York: Columbia University Press, 1992).

35. Raphael Samuel, *Theatres of Memory,* vol. 1, *Past and Present in Contemporary Culture* (London: Verso, 1994), 25.

36. This double take is not unlike the modernist tendency, observed by Bruno Latour in *We Have Never Been Modern* (Cambridge: Harvard University Press, 1993), to simultaneously insist on hybridity and purification—holding on to the ontological division between human and nonhumans (things, machines) while also canceling out their separation. The invincibility of these arguments is possible only because they hold on to the absolute dichotomy between the order of Nature and that of Society, a dichotomy that "is itself only possible because they never consider the work of purification and that of mediation together" (40). On the theory of technology and social change, see Latour's *Aramis, or the Love of Technology* (London: Harvard University Press, 1996).

37. Kuhn, "A Journey through Memory," 183.

38. Steven Rose, *The Making of Memory* (London: Bantam Press, 1992), 91–96.

39. Richard Chalfen, "Snapshots 'R' Us: The Evidentiary Problematic of Home Media," *Visual Studies* 17, no. 2 (2002): 141–49, 144.

40. John Urry, "How Societies Remember the Past," in *Theorizing Museums: Representing Identity and Diversity in a Changing World,* ed. S. Macdonald and G. Fyfe (Oxford: Blackwell, 1996), 45–68.

41. George Lipsitz, *Time Passages: Collective Memory and American Popular Culture* (Minneapolis: University of Minnesota Press, 1990), 5.

42. For a very interesting historical exploration of the notion of time in individual remembering, see the work of Douwe Draaisma, particularly his book *Waarom de tijd sneller gaat als je ouder wordt* (Groningen: Historische Uitgeverij, 2001). For a specific study on the (historical) use of metaphors in relation to memory, see Draaisma, *Metaphors of Memory: A History of Ideas about the Mind* (Cambridge: Cambridge University Press, 2000).

43. I am referring here to George Lakoff and Mark Johnson's renowned theory of metaphors as described in *Metaphors We Live By* (Chicago: Chicago University Press, 1980).

44. John B. Thompson, *The Media and Modernity: A Social Theory of the Media* (Cambridge: Polity Press, 1995).

45. Thompson, in *The Media and Modernity*, is obviously concerned with the double-bind that constitutes the process of self-formation in modernity—where the self is caught between an increasing dependency on media and an increasing need for self-reflexivity to define itself as part of a larger world (233).

46. The obvious theoretical link to Gilles Deleuze's theory on the relation between time, movement, and image comes to mind. I extensively elaborate on this theory in Chapter 6.

47. Besides sight and sound, other sensory perceptions, such as smell or touch, form a trigger for later recall. Few contemporary theorists have stressed the role of the senses in *cultural* memory. See, for instance, C. Nadia Seremetakis, ed., *The*

Senses Still: Perception and Memory as Material Culture in Modernity (Boulder, CO: Westview Press, 1994).

48. Thomas Elsaesser " 'Where Were You When . . . ?' or, 'I Phone, Therefore I Am,' " *PMLA* 118, no. 1 (2003): 120–2, 122.

49. Marita Sturken, in *Tangled Memories: The Vietnam War, the AIDS Epidemic, and the Politics of Remembering* (Berkeley: University of California Press, 1997), defines cultural memory as "the memory that is shared outside the avenues of formal historical discourse yet is entangled with cultural products and imbued with cultural meaning" (3).

50. Sturken, *Tangled Memories*, 5–6.

51. Sturken argues in *Tangled Memories*: "I therefore want to distinguish between cultural memory, personal memory, and official historical discourse. I am not concerned in this book with memories insofar as they remain individual" (3).

52. Alison Landsberg, *Prosthetic Memory: The Transformation of American Remembrance in the Age of Mass Culture* (New York: Columbia University Press, 2004), 21.

53. People build up special relationships with the television programs they tape and the music they rerecord for themselves. For an interesting view on the meaning of self-taping television programs, see Kim Bjarkman, "To Have and To Hold: The Video Collector's Relationship with an Eternal Medium," *Television and New Media* 5, no. 3 (2004): 217–46.

CHAPTER 2

1. Charlie Kaufman, *Eternal Sunshine of the Spotless Mind*, directed by Michel Gondry (Universal City, CA: Focus Features, 2004).

2. The movie's producers have created a mock website of the company Lacuna Inc., available at: http://www.lacunainc.com (accessed December 23, 2006).

3. See John Sutton, *Philosophy and Memory Traces: Descartes to Connectionism* (Cambridge: Cambridge University Press, 1998).

4. For an eminent and thorough historical analysis of metaphors of memory, see Douwe Draaisma, *Metaphors of Memory: A History of Ideas about the Mind* (Cambridge: Cambridge University Press, 2000). The library and archive as metaphors are described in chapter 2, "Memoria: Memory as Writing."

5. See Henri Bergson, *Matter and Memory* (London: George Allen and Unwin, 1911). Originally published in 1896. The original manuscript of Matter and Memory was published in 1896.

6. Bergson, *Matter and Memory*, 197.

7. Of course this view is widely disputed as the age-old distinction of mind and brain remains a hotly contested issue. For a neuroscientific take on this issue, see Antonio Damasio, *Looking for Spinoza: Joy, Sorrow, and the Feeling Brain* (Orlando, FL: Harcourt, 2003), chapter 5.

8. For an illuminating introduction to neuroscientific and genetic "machines" of memory, see Rusiko Bourtchouladze, *Memories Are Made of This: How Memory Works in Humans and Animals* (New York, Columbia University Press, 2002).

9. The metaphor of "hardware" to describe the matter of memory may be as tricky as the book retrieval metaphor. It presumes that the brain is a fixed set of neurons and genes that remains unaltered when "software" is run on its electronically wired system. Yet the living cells a brain is composed of are constantly changing due to external and internal stimuli. Brain and mind work in tandem to produce mental images, moods, and feelings, and they mutually inform their altered states. Rather than deploying the term "hardware," I resort to the metaphor of "mindware," a concept introduced by Andy Clark, in *Mindware: An Introduction to the Philosophy of Cognitive Science* (Oxford: Oxford University Press, 2001), to counter the potential misconception anchored in the computer metaphor.

10. The metaphor of the orchestra describing the mind's functions comes from Antonio Damasio, who coins the image in his book *The Feeling of What Happens: Body and Emotion in the Making of Consciousness* (Orlando, FL: Harcourt, 1999), 216. However, whereas Damasio restricts the use of this metaphor to the neuroscientific aspects of the mind, I expand its meaning to include objects external to the body as well as cultural aspects that affect the mind's memory functions.

11. For psychological studies on autobiographical memory and reminiscence, see, for instance, Susan Bluck and Linda J. Levine, "Reminiscence as Autobiographical Memory: A Catalyst for Reminiscence Theory Development," *Aging and Society* 18 (1998): 185–208. See also Linda J. Levine, "Reconstructing Memories for Emotions," *Journal of Experimental Psychology* 126 (1997): 176–77.

12. Steven Johnson, "The Science of *Eternal Sunshine*: You Can't Erase Your Boyfriend from Your Brain, but the Movie Gets the Rest of It Right," *Slate*, March 22, 2004, available at: http://www.slate.com/id/2097502/ (accessed April 18, 2006).

13. Christopher Nolan, *Memento*, directed by Christopher Nolan (Los Angeles: Newmarket Films, 2000). The film is based on Christopher Nolan's brother Jonathan Nolan's short story *Memento Mori*.

14. See also Steven Johnson, *Mind Wide Open: Your Brain and the Neuroscience of Everyday Life* (New York: Scribner, 2004); Edwin Hutchins, *Cognition in the Wild* (Cambridge, MA: MIT Press, 1996).

15. Damasio, *Looking for Spinoza*, 88–98.

16. Experiments in which subjects were asked to invoke a particularly strong emotional episode from their personal memory confirmed Damasio's theory; certain invoked feelings corresponded to certain changes of neuroactivity spotted in specific regions of the brain. For a description of the experiment, see Damasio, *Looking for Spinoza*, 95–101. The mapping of body states significantly altered during the process of feeling, evidenced by the electrical monitors of positron-emission tomography (PET) scans registering the seismic activity of emotion in all experimental subjects before the actual experience of feeling (sadness, joy) had begun.

17. Strictly speaking, the destruction of physical objects patients bring into Lacuna's office should be redundant, because the internal, mental pictures from which the unpleasant memories derive no longer constitute the link between the brain and the feeling or emotion proper.

18. Don Slater, quoted by Deborah Chambers in *Representing the Family* (London: Sage, 2001), refers to the results of a market research survey in which 39 percent of respondents claimed their family photos to be their most treasured possessions (82).

19. Walter Benjamin, *One-Way Street and Other Writings* (London: Verso, 1979). On Benjamin's writings on memory objects, see also Esther Leslie, "Souvenirs and Forgetting: Walter Benjamin's Memory-Work," in *Material Memories*, ed. Marius Kwint, Christopher Breward, and Jeremy Aynsley (Oxford: Berg, 1999), 107–23.

20. Belinda Barnet, "The Erasure of Technology in Cultural Critique," *Fibreculture* 1 (2003), http://journal.fibreculture.org/ (accessed December 23, 2006).

21. John Sutton, "Porous Memory and the Cognitive Life of Things," in *Prefiguring Cyberculture: An Intellectual History*, ed. Darren Tofts, Annemarie Jonson, and Alessio Cavallara (Cambridge, MA: MIT Press, 2002), 130–41, 138.

22. Clark, *Mindware*, 141.

23. L. H. Martin, *Technologies of the Self: A Seminar with Michel Foucault* (London: Tavistock, 1988), 16.

24. My notion of sociocultural practices finds a middle ground between what sociologist Pierre Bourdieu, in *Outline of a Theory of Practice* (Cambridge: Cambridge University Press, 1977), refers to as "habitus" and what philosopher Michel de Certeau, in *The Practice of Everyday Life* (Berkeley: University of California Press, 1984), rearticulates as "the practice of everyday life." Bourdieu's "habitus" is associated with the internalized, practical knowledge by which people operate in stable, social structures and situations; De Certeau uses the term "practice" to emphasize the dynamics of people evolving in social structures, changing them and adapting to new ones. When I use the term "sociocultural practices," I am referring to both static structures and dynamic changes. However, I am much more specific in my denotation of the word "practice," referring to a set of practical, technical, social, and cultural skills needed to operate the "technologies of self" Foucault identifies. These sociocultural practices are grounded both in materiality and technology (in this case media technologies) as well as in the knowledge of their practical use (e.g., social norms and discourses).

25. Hartmut Winkler, a German media scholar, presents a theory of cultural continuity by explaining the translation of certain cultural practices into "deposits" (defined by technology and its use) that turn back into practices. Through constant reinterpretation and reshaping of practices and objects, the continuity of culture is secured, even if constantly morphing. See "Discourses, Schemata, Technology, Monuments: Outline for a Theory of Cultural continuity," *Configurations*, 10 (2002): 91–109.

26. Anthropologist Edwin Hutchins, in *Cognition in the Wild* (Cambridge, MA: MIT Press, 1996), argues in contrast to the standard view that culture affects the cognition of individuals, that cultural activity systems have cognitive properties of their own that are different from the cognitive properties of the individuals who participate in them.

27. Roger Silverstone, Eric Hirsch, and David Morley, "Information and Communication Technologies and the Moral Economy of the Household," in *Consuming Technologies: Media and Information in Domestic Spaces,* ed. Roger Silverstone and Eric Hirsch (London: Routledge, 1992), 14–31.

28. Notions of self and family, as I argue Chapter 6, are constructed and reflected through media technologies. Media technologies, as Silverstone, Hirsch, and Morley argue in "Information and Communication Technologies," are never fixed instruments, just as media objects are never immutable items. Video cameras may be appropriated differently by various members of a household, and it is not uncommon that each member of a household composes his or her own individual photo album in addition to the family album kept by a parent.

29. Positron-emission tomography (PET) is a scanning technology that with the help of radioactive isotopes allows one to study the brain functions in vivo; functional magnetic resonance imaging (fMRI) makes it possible to record static images of activity in the brain that subsequently can be turned into a moving film.

30. With the digitization of medical diagnostics came a stronger articulation of images as transparent indicators of ailments, even though it has been abundantly argued that (medical) imaging has rendered the body opaque rather than transparent. For an elaboration of this argument, see José van Dijck, *The Transparent Body: A Cultural Analysis of Medical Imaging* (Seattle: University of Washington Press, 2005), chapter 1.

31. For an insightful analysis of how brain images like those from PET scans have served in courts and popular culture as "objective" evidence of mental illness and abnormality, see Joseph Dumit, "Objective Brains, Prejudicial Images," *Science in Context* 12, no. 1 (1999): 173–201. See also Brent Garland, ed., *Neuroscience and the Law: Brain, Mind, and the Scales of Justice* (New York: Dana Press, 2004).

32. Neurologists' and neuroscientists' infatuation with fMRI as a way to determine pathological and criminal behavior is also touted as the new "phrenology" of medicine; see William R. Uttal, *The New Phrenology* (Cambridge, MA: MIT Press, 2003).

33. Many science fiction movies, from *The Matrix* to *The Thirteenth Floor,* prophesy the future of human bodies to be one where uploading the mind into the computer helps transcend the flesh, ushering into a kind of wired universe where the mind-machine survives autonomously. The merger of brain and computer implicitly hails the triumph of informatics over flesh, of software and hardware over "wetware." N. Katherine Hayles, in *How We Became Posthuman: Virtual Bodies in Cybernetics, Literature and Informatics* (Chicago: University of

Chicago Press, 1999), rightly criticizes theorists such as Hans Moravec and Ray Kurzweil whose affection for "disembodied minds" and "virtual brains" seems to dispose of the body as a locus for mental activity. "Posthumanists," as Hayles calls them, are blind to the materiality of informatics and indifferent to the embodiment of digital media. The idea of human memory being digitized and transposed to a locus outside the brain is an immensely popular trope in the twenty-first century, informing both visionary science projects and science fiction movies like *Brain Destroyer* and *Fantastic Voyage II: Destination Brain.*

34. Eugene Thacker, "What is Biomedia?" *Configurations* 11 (2003): 47–79.

35. Ibid., 76–77.

36. Genomics is a case in point: the computations of genetic sequences are carried out by computers, and thus digital information becomes an impetus for redressing our knowledge of genetic defects. For a detailed explanation of how genomics and information interact, see José van Dijck, *ImagEnation: Popular Images of Genetics* (New York: New York University Press, 1998), chapter 6.

37. For an extensive analysis of how ultrasound imaging not just works to reconfigure our conceptualization of the fetus, but also affects pregnancy and the development of the fetus, see Van Dijck, "Ultrasound and the Visible Fetus," in *The Transparent Body,* 100–117.

38. There is a hilarious scene in *Eternal Sunshine* where Dr. Mierzwiak asks Joel to unleash his painful memories of Clementine by talking about her into the microphone of an old-fashioned cassette recorder. Later in the movie, Joel and Clementine are confronted with their embarrassing monologues when the magnetic tapes with their voices are returned to them through a disgruntled, revengeful secretary after she has discovered the "erased" love affair with her boss, Dr. Mierzwiak.

39. Gregory Ulmer, for instance, treats (digital) memory as a reservoir for creative invention and intervention—new media technologies allowing the reordering and reshaping of digital imprints of the past, whether pictures, sounds, or texts. See Gregory Ulmer, *Heuretics: The Logic of Invention* (Baltimore, MD: Johns Hopkins University Press, 1994).

40. A concise and insightful article in the *New York Times* provides an overview of the many problems involved in storing, preserving, and retrieving digital memory files for the next generation. See Katie Hafner, "Even Digital Memories Can Fade," *New York Times,* November 10, 2004 (online edition, www.nytimes.com).

41. See Jay Bolter and Richard Grusin, *Remediation: Understanding New Media* (Cambridge, MA: MIT Press, 1999).

CHAPTER 3

1. Susan Herring, an American sociologist from Indiana University specializing in computer-mediated communication, in a 2005 presentation, quotes the number

of weblog users from the statistics of the Perseus group at 4.12 million. This number of bloggers also includes hosted weblog services; 34 percent of these logs are used actively. Herring, "Weblog as Genre, Weblog as Sociability," available at: http://ella.slis.indiana.edu/~herring/ssc.ppt (accessed April 19, 2006).

2. Initially, blogs were either personal homepages operated by individuals, mostly people who were interested in sharing technical and personal knowledge, or they were websites consisting of chronological lists of links, interspersed with information and editorialized and personal asides. For a description of the early development of weblogs, see, for instance, *We've Got Blog: How Weblogs Are Changing Our Culture,* ed. John Rodzvilla (Cambridge, MA: Perseus Publishers, 2002); see also Charles Cheung, "A Home on the Web: Presentations of Self on Personal Homepages," in *Web Studies: Rewiring Media Studies for the Digital Age,* ed. David Gauntlett (London: Arnold, 2000), 43–51.

3. Researchers of the blogosphere distinguish lifelogs from linklogs; linkloggers primarily post links to other websites, whereas lifeloggers primarily post details about their personal lives and everyday experiences. See Frank Schaap, "Links, Lives, Logs: Presentation in the Dutch Blogosphere," in *Into the Blogoshphere: Rhetoric, Community and Culture of Weblogs,* ed. Laura Gurak, Smiljana Antonijevio, Laurie Johnson, Clanoy Ratliff, and Jessica Reyman, (Minneapolis: University of Minnesota, 2004), http://blog.lib.umn.edu/blogosphere/ (accessed April 18, 2006).

4. German media theorist Andreas Kitzmann, in his article "That Different Place: Documenting the Self within Online Environments," *Biography* 26, no. 1 (Winter 2003): 48–65, proposes to study media change in the context of the much wider phenomenon of "material complexification" to understand the continuities and changes between old and new media such as diaries and weblogs. He argues that change is not cumulative "but [measured by] structural shifts that may lead to growth, contraction, stasis, or a combination of all three" (51).

5. For an extensive elaboration of this theory, see Antonio Damasio, *The Feeling of What Happens: Body and Emotion in the Making of Consciousness* (Orlando, FL: Harcourt, 1999), 308, chapter 2.

6. Silvan Tomkins, *Affect, Imagery, Consciousness* (New York: Springer, 1962), quoted by Anna Gibbs in "Contagious Feelings: Pauline Hanson and the Epidemiology of Affect," *Australian Humanities Review* 24 (2001). Available at: http://www.lib.latrobe.edu.au/AHR/archive/Issue-December-2001/gibbs.html (accessed April 18, 2006).

7. See Gibbs, "Contagious Feelings."

8. The diary genre has been defined as therapy or self-help, as a means of confession, as a chronicle of adventurous journeys (both spiritual and physical), and as a scrapbook for creative endeavors. Thomas Mallon, author of the standard work, *A Book of One's Own: People and Their Diaries* (New York: Ticknor and Fields, 1984), distinguishes at least seven types of diaries and labels the various

types according to their author's profession or character: chroniclers, travelers, creators, confessors, and so on. Philip Lejeune, in *Le pacte autobiographique* (Paris: Editions du Seuil, 1993), inventories various types of autobiographical writing (diary, letters, autobiography) by their "morphological" features, whereas French literary theorist, Beatrice Didier, in *Le Journal Intime* (Paris: Editions du Seuil, 1976), articulates a more general classification, based on the content of entries, between the personal or private diary (*le journal intime*) and the more public or factual journal. Yet another French literary scholar, Eric Marty, in *L'ecriture du jour: Le journal d'André Gide* (Paris: Editions du Seuil, 1985), classifies diaries by their addressees: are they strictly secret or also written for others?

9. The lifelog is as polymorphous as its paper precursor, and yet when researching new functions and forms of the diary in the digital era, the old typology in terms of content and directionality continues to inform the epistemology of the lifelog. For instance, a Japanese study of weblogs departs from notion that they can be classified according to their contents as "records of fact" or "expression of sentiment" or according to their directionality as "written for oneself" or "written for others." See Yasuyuki Kawaura, Yoshiro Kawakami, and Kiyomo Yamashita, "Keeping a Diary in Cyberspace," *Japanese Psychological Research* 40, no. 4 (1998): 234–45. Many studies classify weblogs along binary axes of self and others, or of personal and public, even though some recent studies tend to acknowledge how the multiple ancestry of weblogs precludes such dichotomous classification.

10. I thank Janelle Taylor for pointing me toward several Alzheimer's patients' lifelogs.

11. One of the initiators of Alzheimer's patients' blogging efforts, Friedell was diagnosed with AD in September 1998. A retired professor of sociology at University of California–Santa Barbara, he began writing and publishing about his symptoms on his personal webpage, available at: http://members.aol.com/MorrisFF (accessed April 18, 2006). DASNI is an Internet-based dementia advocacy and support group whose members share and exchange their experiences and tested remedies. Friedell's weblog is linked to a number of patients' weblogs.

12. Morris Friedell extensively describes this discussion, as the scans of his brain offer anything but a conclusive diagnosis. Available at http://members.aol .com/morrisff/Diagnosis.html (accessed December 23, 2006). He refers to a 2004 *Newsday* article quoting Dr. Ronald Petersen, a researcher at the Mayo Clinic: "You can't take any individual scan and say this person has Alzheimer's. We have an ethical and moral obligation not to cause undue worry or even a misdiagnosis. The technology is evolving, but we're not there yet."

13. Anita Blanchard, "Blogs as Virtual Communities: Identifying a Sense of Community in the Julie/Julia Project," *Into the Blogoshphere: Rhetoric, Community and Culture of Weblogs,* ed. Laura Gurak, Smiljana Antonijevio, Laurie Johnson, Clanoy Ratliff, and Jessica Reyman (Minneapolis: University of Minnesota, 2004), http://blog.lib.umn.edu/blogosphere/ (accessed April 18, 2006).

14. Chip Gerber, "My Journey," March 2005, part 2, available at: http://www.zarcrom.com/users/alzheimers/chip.html (accessed April 18, 2006).

15. See Mary Lockhart's blog, available at: http://www.angelfire.com/ok4/mari5113/index.html (accessed April 18, 2006).

16. Robert Payne, "Digital Memories, Analogues of Affect," *Scan: Journal of Media Arts Culture* 2 (2004), http://scan.net.au/scan/journal/display.php?journal_id=42 (accessed April 18, 2006).

17. The term "community of records" is used by Eric Ketelaar in the context of collective archiving. See Eric Ketelaar, "Sharing: Collected Memories in Communities of Records," *Archives and Manuscripts* 33 (2005): 44–61.

18. Gerber, "My Journey," March 2005, part 3.

19. See Friedell's weblog, available at: http://members.aol.com/MorrisFF?.

20. For a detailed description of the authenticity debate concerning Anne Frank's manuscript in contrast to the fabricated Adolf Hitler diaries that were discovered in the 1980s, see Sonja Neef: "Authentic Events: The Diaries of Anne Frank and the Alleged Diaries of Adolf Hitler," in *Sign Here! Handwriting in the Age of Technological Reproduction*, ed. Sonja Neef, José van Dijck, and Eric Ketelaar (Amsterdam: Amsterdam University Press, 2006), 23–50.

21. As Canadian archivist Jane Zhang claims in "The Lingering of Handwritten Records," *Proceedings of I-Chora Conference* (International Conference on the History of Records and Archives, University of Toronto, October 2–4, 2003), 38–45, an individual's handwriting is habitually viewed as "his own personal mark, which distinguishes him not only from others, but also from his own past and future" (43).

22. Sonja Neef, "Die (rechte) Schrift und die (linke) Hand," *Kodikas/Ars Semiotica* 25, no. 1 (2002): 159–76.

23. Jacques Derrida, *Archive Fever: A Freudian Impression* (Chicago: University of Chicago Press, 1995).

24. Friedrich Kittler, in his famed *Film, Gramophone, Typewriter* (Stanford, CA: Stanford University Press, 1999), states that the typewriter disrupted the intimacy of handwritten expression, as it "tears writing from the essential realm of the hand, i.e., the realm of the word" (198). It should be noted, though, that this idea does not originally stem from Kittler; he is referring to Martin Heidegger's Parmenides lecture.

25. For an intricate philosophical explanation of how technology and body are intertwined in the digital act of writing, see Mark Hansen, *Embodying Technesis: Technology beyond Writing* (Ann Arbor: University of Michigan Press, 2000).

26. Emily Nussbaum, in "My So-Called Blog," *New York Times Magazine*, January 11, 2004, online version at (www.nytimes.com), states: "A LiveJournal is a Blurty is a Xanga is a DiaryLand."

27. However, there is another side to this techno-cultural transformation that often is underemphasized. The culture of reciprocation is not solely based on linking the self to the Internet, but it is also based on linking the Internet to the self. Tracing

cultural or political preferences of other bloggers, one can decide to connect to people with similar tastes and preferences; it is precisely this technological feature that makes weblogs interesting for marketers. With the use of fairly simple software applications like AllConsuming.net, it becomes increasingly easy to find correlations between bloggers and the cultural products they mention via links or sidebars: books, music, television programs, movies, and so forth. On the interlinking of weblogs for commercial purposes, see Erik Benson, "All Consuming Web Services," O'Reilly Webservices, May 7, 2003, http://www.xml.com/pub/a/ws/2003/05/27/allconsuming.html (accessed April 18, 2006).

28. Tracking software allows a glimpse of the patterns and trends that emerge out of the topics shared by a group. Coupled with vast databases like those of Amazon and Google, the possibilities for polling and marketing research are endless, explaining Google's eagerness to buy start-up companies like Blogger. Google bought Pyra labs, one of the first start-up companies that designed blogger software, in 2003. For details on this transaction, see Leander Kahney, "Why Did Google Want Blogger?" *Wired News*, February 22, 2003, http://www.wired.com/news/technology.

29. Mallon, *A Book of One's Own*, xvi.

30. On the importance of the addressee in diaries, see Marty, *L'ecriture du jour*, 87.

31. William M. Decker, in *Epistolary Practices: Letter Writing in America before Telecommunications* (Raleigh: University of North Carolina Press, 1998), theorizes the evolution of epistolary writing in the United States; he observes that letters, much like diaries, carry the aura of a private genre, whereas the diary genre encodes itself according to public standards: "What we identify as the private life is a conventionalized and hence public construction" (6).

32. Elizabeth Yakel, in "Reading, Reporting, and Remembering: A Case Study of the Maryknoll Sisters Diaries," *Proceedings of I-Chora Conference* (International Conference on the History of Records and Archives, University of Toronto, October 2–4, 2003), 142–50, describes an intriguing account of how the Maryknoll Sisters, a religious community active between 1912 and 1967, adapted the genre as a collective means of expression to record and exchange spiritual and intellectual journeys to each other. Their record-keeping practices suited various goals, from expressing individual beliefs to communicating information across time and space with like-minded congregations. As Yakel phrases it: "The diaries had multiple audiences—they were a means of internal communication within the community and also served as a mechanism for external communication to Catholics and others interested in their mission activities" (143).

33. Michael Piggott, archivist at the University of Melbourne, discovered Australian archives to contain many such collective ego-documents, chronicling important episodes from the sixteenth to the nineteenth centuries through the eyes of transient groups. See Michael Piggott, "Towards a History of the Australian Diary," *Proceedings of I-Chora Conference* (International Conference on

the History of Records and Archives, University of Toronto, October 2–4, 2003), 68–75.

34. As some theorists have explained, the construction of self on the Internet often results in various forms of self-glorification or narcissism, in addition to creating multiple identities and deliberately testing those in the partial anonymous waters of online blogging communities. See, for instance, Joanne Jacobs, "Communication over Exposure: The Rise of Blogs as a Product of Cybervoyeurism" (paper presented at the ANZCA conference at Brisbane, "Designing Communication for Diversity," July 2003), available at: http://www.joannejacobs.net/pubs.html (accessed December 23, 2006).

35. As Australian sociologist David Chaney observes in *Cultural Change and Everyday Life* (Houndmills, UK: Palgrave, 2002), everyday life is a creative project "because although it has the predictability of mundane expectations, it is simultaneously being worked at both in the doing and in retrospective reconsideration" (52).

36. Jan Fernback, in "Legends on the Net: An Examination of Computer-Mediated Communication as a Locus of Oral Culture," *New Media and Society* 5, no.1 (2003): 29–45, remarks that "as mediated human communication becomes more and more non-linear, decentralized, and rooted in multimedia, the distinction between orality and literacy becomes less evident and less important" (29).

37. Nussbaum, "My So-Called Blog."

38. Susan Herring provides this statistic in her powerpoint presentation (see note 1), available at: http://ella.slis.indiana.edu/~herring/ssc.ppt.

39. Esther Milne, "Email and Epistolary Technologies: Presence, Intimacy, Disembodiment," *Fibreculture* 2 (2004), http://journal.fibreculture.org/issue2/issue2_milne.html (accessed April 18, 2006).

40. Fernanda Viegas, "Blog Survey: Expectations of Privacy and Accountability," MIT Media Lab Survey, 2004, http://web.media.mit.edu/~fviegas/survey/blog/results.htm (accessed April 18, 2006).

41. It may be instructive to compare blogs and blogging to the use of the mobile phone. Alex Taylor and Richard Harper, in their study of teenagers' use of cell phones, "The Gift of the Gab? A Design-Oriented Sociology of Young People's Use of Mobiles," *Journal of Computer Supported Cooperative Work* 12, no. 3 (2003): 267–96, note how phone-mediated activities resemble established social practices such as gift giving; the ritual of gift exchange is now extended to symbolic messages (SMS or spoken), and, like the material equivalent, it is rooted in a mental scheme of obligation and reciprocation. Through a subtle system of shared norms for exchanging phones, rationing access to personal messages, and obligations to respond, users assign symbolic value to tangible or virtual objects.

42. In their illuminating analysis of the phenomenon, sociologists Miller and Shepherd argue that blogging should be regarded as social action—a "new rhetorical opportunity" that needs to be examined in terms of its use. See Carolyn Miller and Dawn Shepherd. "Blogging as Social Action: A Genre Analysis of the Weblog,"

in *Into the Blogoshphere: Rhetoric, Community and Culture of Weblogs*, ed. Laura Gurak, Smiljana Antonijevio, Laurie Johnson, Clanoy Ratliff, and Jessica Reyman (Minneapolis: University of Minnesota, 2004). Available at: http://blog.lib.umn .edu/blogosphere/ (accessed April 18, 2006).

43. In the case of phone conversations and text messaging, Taylor and Harper, in "The Gift of the Gab?" found that some teenagers express the wish to store each SMS exchange on a memory card in order to recall the experience later: the message's physical properties (form, content, and time and date stamp) all work in combination to instill meaning onto the physical object.

44. According to Viegas's "Blog Survey," almost 75 percent of all bloggers indeed edit their past entries, with changes varying from punctuation and grammar to contents and names.

45. Nussbaum notes in "My So-Called Blog" that bloggers have a "degraded or relaxed sense of privacy," depending on your perspective: "Their experiences may be personal, but there is no shame in sharing . . . [and they get back] a new kind of intimacy, a sense that they are known and listened to."

46. Not surprisingly, more than one-third of all bloggers have gotten into trouble because of things they have written in their blogs and the majority forgets about defamation or liability when writing about others in networked environments. As the aforementioned MIT Media Lab Survey by Viegas shows, bloggers are hardly concerned with the persistent nature of what they publish; the overwhelming majority of them publish private information about themselves or other people without thinking about legal or moral consequences.

CHAPTER 4

1. There have been a number of psychological and cognitive studies of the connection between emotion and individual meanings attached to music, and several of the more important ones are surveyed in the course of this chapter. One of the oldest and often cited studies in this respect is Leonard B. Meyer, *Emotion and Meaning in Music* (Chicago: Chicago University Press, 1961).

2. The list of studies on popular music and cultural memory and identity is remarkably long. Most studies on collective memory and identity cover recorded music as part of popular culture. See, for instance, George Lipsitz, *Time Passages: Collective Memory and American Popular Culture* (Minneapolis: University of Minnesota Press, 1990). For an extensive bibliography and an overview of academic work on popular music and collective identity, see John Connell and Chris Gibson, *Sound Tracks: Popular Music, Identity, and Place* (New York: Routledge, 2003).

3. This chapter concentrates on popular (rather than classical or experimental-artistic) music and its affective commitment to memory, because pop music is probably more conducive to the kind of mental mapping and narrative recall fundamental to the argument developed here. It is beyond the scope of this chapter to

theorize how these processes would extend to other kinds of music, which may create a different connection to identity and memory.

4. See Jeffrey Prager, *Presenting the Past: Psychoanalysis and the Sociology of Misremembering* (Cambridge, MA: Harvard University Press, 1998), 215.

5. Timothy Taylor, in *Strange Sounds: Music, Technology and Culture* (New York: Routledge, 2001), puts more emphasis on the sociotechnical systems from which recorded music emanates and how this becomes part of its history and collective memory. See also David Morton, *Off the Record: The Technology and Culture of Sound Recording in America* (New Brunswick, NJ: Rutgers University Press, 2000). Morton defines recording culture as follows: "The recording of music is an activity that combines a very old form of cultures, the performance of music, with a variety of technological processes to create a new form of culture. . . . What we are concerned with here is not only music captured on record as an example of mass-produced culture but also recording as a cultural process; not only the meaning of the content of a record, but the meaning of the practices which developed around the act of recording" (13).

6. Started as a onetime millennium event in 1999, the Dutch national public radio station (Radio 2) invited listeners to send in their personal top-five favorite songs of all time, resulting in a collective Top 2000. (Available at: http://top2000 .radio2.nl/2005/site/page/homepage.) The response to this event was so overwhelming that the station decided to repeat it the next year, and a tradition began. In December 2004 and 2005, the national Top 2000 was selected by well over one million Dutch citizens who sent in their personal top-five songs. The number of public participants is unprecedented in the history of mediated events in The Netherlands. In 2004, almost 6.5 million people listened to the radio broadcast, 5 million people watched the accompanying daily television shows, and the website registered 9.2 million page views in just five days. Cast against a population of 16 million, the event engaged more than half of all Dutch people twelve years and older. The comments used in this chapter are derived from the 2004 database; this database is no longer publicly available but is archived by Radio 2. Comments were originally in Dutch; they were translated by me and I have identified the respondents in the same way they identified themselves on the (public) website. I thank Kees Toering, station manager and initiator of the Top 2000 for making all statistics and archives available to me.

7. Music turns out to be an important contextual element in human recall; clinical research shows that recall is optimal when people hear the same music during the experiencing and recalling of events. See W. R. Balch and B. S. Lewis, "Threads of Music in the Tapestry of Memory," *Memory and Cognition* 21 (1996): 21–8.

8. Thomas Turino, in "Signs of Imagination, Identity, and Experience: A Peircian Semiotic Theory for Music," *Ethnomusicology* 43, no. 2 (1999): 221–55, argues that identities are at once individual and social and that music is a key resource for realizing personal and collective identities at the same time.

9. Patrick Colm Hogan, *Cognitive Science, Literature, and the Arts* (New York: Routledge, 2003), 14.

10. Friedrich Kittler, in *Gramophone, Film, Typewriter* (Stanford, CA: Stanford University Press, 1999) describes the historical tendency to regard the gramophone as an instrument of repetition and faithfulness; Freud took this very literally by explaining how "the unconscious coincides with electric oscillations" (89).

11. Lisa Gitelman, "How Users Define New Media: A History of the Amusement Phonograph," in *Rethinking Media Change: The Aesthetics of Transition,* ed. David Thorburn and Henry Jenkins (Cambridge, MA: MIT Press), 61–80, 65.

12. American neuroscientist Rusiko Bourtchouladze argues in *Memories Are Made of This: How Memory Works in Humans and Animals* (New York: Columbia University Press, 2002) that all memories start as episodic, but only unique experiences survive as time goes by. "Those that do not have freshness and characteristic flavor tend to go downhill with time" (28).

13. For a detailed explanation of this argument, see H. Baumgartner, "Remembrances of Things Past: Music, Autobiographical Memory, and Emotion," *Advances in Consumer Research* 19 (1992): 613–20.

14. I have tried to paraphrase the points Antonio Damasio makes about emotions and feelings with regard to memory in *The Feeling of What Happens: Body and Emotion in the Making of Consciousness* (Orlando, FL: Harcourt, 1999), 183–94.

15. See Damasio, *The Feeling of What Happens,* 188.

16. See Turino, "Signs of Imagination, Identity, and Experience," 224.

17. See also Simon Frith, who, in *Performing Rites: On the Value of Popular Music* (Oxford: Oxford University Press, 1996), observes that music, especially during adolescence and teenage years, tends to be retained in connection to intense personal experiences; particular recordings are often considered "our songs" by a group or collective focused on identity building and enhancement.

18. See Damasio, *The Feeling of What Happens,* 123–24.

19. A few cognitive psychological studies have shown how older adults' memory grows more positive over the years. As Quinn Kennedy, Mara Mather, and Laura Carstensen show in "The Role of Motivations in the Age–Related Positivity Effect in Autobiographical Memory," *Psychological Science* 15, no. 3 (1994): 208–14, older adults are more motivated than younger adults to remember their past in emotionally satisfying ways, and older adults' positive bias in reconstructive memory reflects their motivation to regulate emotional experience.

20. Geoffrey O'Brien, *Sonata for Jukebox: Pop Music, Memory, and the Imagined Life* (New York: Counterpoint, 2004), 16.

21. American clinical psychologists Matthew Schulkind, Laura Hennis, and David Rubin, in "Music, Emotion and Autobiographical Memory: They're Playing Your Song," *Memory and Cognition* 27, no. 6 (1999): 948–55, tested how various age groups remember through music. For their experiment, the researchers tested two

groups of adults: younger adults between eighteen and twenty-one years of age and older adults between sixty-five and seventy years of age. They made them listen to a series of songs that were popular between 1935 and 1994 but only appeared on the hit lists during a defined period (in contrast to evergreens). The subjects were asked whether each song reminded them of a general period or a specific event from their lives.

22. Connell and Gibson, *Sound Tracks*, 222–23.

23. At this point, I will not enter the discussion whether this intergenerational longing concerns a generation or a specific time frame, such as a decade. Joseph Kotarba, in "Rock 'n' Roll Music as a Timepiece," *Symbolic Interaction* 25, no. 3 (2002): 397–404, argues that the concept of cohort is more useful than the concept of the decade for an interpretive analysis of musical reminiscence.

24. See Antonio Damasio, *Looking for Spinoza. Joy, Sorrow and the Feeling Brain* (Orlando, FL: Harcourt, 2003), 93–96. Also see the more detailed description in Chapter 2.

25. Barbra Misztal, in *Theories of Social Remembering* (Maidenhead UK: Open University Press, 2003) identifies this as generational memory: "As generation follows generation, each receives an inheritance from its predecessor, and this intergenerational transmission, or tradition, is a foundation of societal continuity" (84).

26. Mark Katz, in *Capturing Sound: How Technology Has Changed Music* (Berkeley: University of California Press, 2004) and Eric W. Rothenbuhler and John Durham Peters, in "Defining Phonography: An Experiment in Theory," *Musical Quarterly* 81, no. 2 (1997): 242–64, all discuss the significance of (vinyl) materiality in the age of phonography in contrast to the pre-phonographic and post-phonographic era. Like Katz, Rothenbuhler and Peters emphasize the important role of technology in the history of recorded music.

27. On the material temporality of recording, see the discussion on "triple temporality" in chapter 6 of Jonathan Sterne's *The Audible Past* (Durham, NC: Duke University Press, 2003).

28. In chapter 4 of his book *Television: Technology and Cultural Form* (Hanover, NH: University Press of New England, 1974), Raymond Williams explains the term "private mobilization."

29. Joe Tacchi, in his article "Radio Texture: Between Self and Others," in *Material Cultures: Why Some Things Matter*, ed. Daniel Miller (London: UCL Press, 1998), 25–45, points at how hearing favorite songs on the radio is different than playing them on your own stereo. Hearing a song on the radio, as Eric W. Rothenbuhler argues in "Commercial Radio as Communication," *Journal of Communication* 46 (1996): 125–43, is a moment when the symbolic activity of the other enters into the field of contingencies of the self.

30. Joseph Auner, "Making Old Machines Speak: Images of Technology in Recent Music," *ECHO: A Music-Centered Journal* 2, no. 2 (2000), http://www.humnet .ucla.edu/echo (accessed April 19, 2006).

31. At the earliest stages of digitization, Alan Goodwin, in "Sample and Hold: Pop Music in the Digital Age of Reproduction," *Critical Quarterly* 30, no. 3 (1988): 34–49, already argued for a new postmodernist theory of musical creativity, based upon the new digitally based cultural practice of sampling. However, the politics and aesthetics of sampling fall outside the boundaries of this chapter.

32. Katz, *Capturing Sound*, 171.

33. Tia de Nora, *Music in Everyday Life* (Cambridge: Cambridge University Press, 2000), 46–74.

34. See, for instance, Simon Frith, *Performing Rites: On the Value of Popular Music* (Oxford: Oxford University Press, 1996).

35. Michael Bull, *Sounding Out the City: Personal Stereos and the Management of Everyday Life* (Oxford: Berg 2000), 40.

36. Websites offering mix-and-burn software (such as iTunes, Blaze Audio, The Music Tablet) display a tendency to address users as "interactive creators" who "transform music buying into an instant creative experience." Art of the Mix, a website that promotes the swapping of mixed CDs, lists a large number of personal motivations for creating playlists and burning them onto a CD, including "the romantic mix, the break-up mix, the hangover mix, the airplane mix, and the sick-in-bed mix," to match fleeting moods and personal circumstances. See Art of the Mix, available at: http://www.artofthemix.org/writings/history.asp (accessed April 19, 2006).

37. A number of studies using undergraduates as research objects point out how current mood influences memory. See, for instance, Gordon Bower and Joseph Forgas's article "Affect, Memory, and Social Cognition," in *Cognition and Emotion*, ed. Joseph Forgas, Eric Eich, and Gordon Bower (Oxford: Oxford University Press, 2000), 87–168. On the relation between age and (positive) remembering, Mara Mather, in "Aging and Emotional Memory," in *Memory and Emotion*, ed. Daniel Reisberg and Paula Hertel (Oxford: Oxford University Press, 2004), 272–307, concludes that among younger adults, negative mood increases the likelihood of remembering negative information.

38. Jonathan Sterne contends in his article "MP3 as Cultural Artifact," *New Media and Society*, 8, 5 (2006): 825–42, that the MP3 is a cultural artifact in its own right; in line with Sterne's thinking, I argue that music recorded on digital files and played by MP3 players triggers sociocultural practices that deserve to be regarded in their own right rather than as a derivative of earlier portable technologies.

39. On the significance of continuation and change in historical transformations of sound technologies and their accompanying collective sociocultural practices, see Lisa Gitelman, "How Users Define New Media."

40. William H. Kenney, *Recorded Music in American Life: The Phonograph and Popular Memory, 1890–1945* (Oxford: Oxford University Press, 1999), xix.

41. In December 2005, a national discussion erupted in newspapers around the country when listeners voted a new number-one song to top the ranking—"Avond,"

a song in Dutch, composed and sung by Dutch artist Boudewijn de Groot—thus defeating the longtime English number-one song (Queen's "Bohemian Rhapsody").

42. Paul Grainge, in his article "Nostalgia and Style in Retro America: Moods, Modes and Media Recycling," *Journal of American and Comparative Cultures* 23, no. 1 (2000): 27–34, considers nostalgia for music of the past as a "socio-cultural response to forms of discontinuity, claiming a vision of stability and authenticity in some conceptual 'golden age' " (28).

43. See Lawrence Lessig, *The Future of Ideas: The Fate of the Commons in a Connected World* (New York: Vintage, 2002).

44. Quite a few clones of the Top 2000 have been initiated by competing commercial stations in The Netherlands, but none of these imitations gather nearly as much clout as the public event. None of these clones have a comparable investment in public participation and exchange via websites and chat forums.

CHAPTER 5

1. I prefer the term "personal photography" over commonly used terms like "amateur photography" or "family photography." The word "personal" is meant to distinguish it from professional photography, but it also avoids the troubling connotation of "amateurish" in relation to camera use. Family photography mistakenly presupposes the presence of a familial context, whereas photography has always been and is increasingly used for personal identity formation.

2. Norwegian researcher C. M. Stuhlmiller, in "Narrative Picturing: Ushering Experiential Recall," *Nursing Inquiry* 3 (1996): 183–84, argues that narratives of remembering always involve elements of imagining and picturing, feeding verbal stories.

3. Susan Sontag's *On Photography* (New York: Delta, 1973) and Roland Barthes's *Camera Lucida: Reflections on Photography* (New York: Hill and Wang, 1981) were not the first but were certainly the most notable theories of photography. Both essayists claim memory to be the most important function of personal photography, but they also acknowledge photography's material, ritual, and communicative meaning in the everyday lives of people.

4. "My pictures are my memories" is a cliché still resonant in many anthropological and sociological studies of family photography. See, for instance, Richard Chalfen, "Snapshots 'R' Us: The Evidentiary Problematic of Home Media," *Visual Studies* 17, no. 2 (2002): 141–49.

5. Barthes, *Camera Lucida*, 80.

6. According to neurobiologist Steven Rose in *The Making of Memory* (London: Bantam Press, 1992), the still photograph is not a fixed object, but it forms the input of a constantly changing and evolving autobiographical memory.

7. For a very insightful overview of research in this area, see Deryn Strange, Matthew Gerrie, and Maryanne Garry, "A Few Seemingly Harmless Routes to a False Memory," *Cognitive Process* 6 (2005): 237–42.

8. There are a large number of research groups reporting on the issue of false memory as it is related to both narrative and visual evidence. For instance, see Helene Intraub and James Hoffman, "Reading and Visual Memory: Remembering Scenes that Were Never Seen," *American Journal of Psychology* 105, no. 1 (1992): 101–14. See also Elisabeth Loftus, "The Reality of Repressed Memories," *American Psychologist* 48 (1993): 508–37; Elisabeth Loftus and J. Pickrell, "The Formation of False Memories," *Psychiatric Annals* 25 (1995): 720–25. On the role of true pictures in the creation of false memories, see Stephen Lindsay, Lisa Hagen, Don Read, Kimberley Wade, and Maryanne Garry, "True Photographs and False Memories," *Psychological Science* 15, no. 3 (2004): 149.

9. Research by cognitive psychologists focusing particularly on the role of doctored photographs in relation to false memory is also widely available. See, for instance, Maryanne Garry and Matthew Gerrie, "When Photographs Create False Memories," *Current Directions in Psychological Science* 14 (2005): 321. See also Kimberley Wade, Maryanne Garry, Don Read, and Stephen Lindsay, "A Picture Is Worth a Thousand Lies: Using False Photographs to Create False Childhood Memories," *Psychonomic Bulletin and Review* 9, no. 3 (2002): 597–603. In their article, Wade and her colleagues confronted experimental subjects with doctored pictures of the subjects as children in hot-air balloons; half the adults were persuaded into believing they actually remembered the balloon ride.

10. For the ongoing debate on whether narratives or pictures are more conducive to false memories, see Maryanne Garry and Kimberley Wade, "Actually, a Picture Is Worth Less Than 45 Words: Narratives Produce More False Memories Than Photographs Do," *Psychonomic Bulletin and Review* 12, no. 2 (2005): 359–66.

11. See, for instance, Gerald Zaltman, *How Customers Think: Essential Insights into the Mind of the Market* (Boston: Harvard Business School Press, 2003). An experiment has shown how easy it is for advertisements to plant false memories: researchers presented participants with fake ads for Disney that feature them meeting Bugs Bunny at a Disney Resort (although Bugs Bunny is a Warner Brothers character). Most participants believed this was a true experience. See Kathryn Braun, Rhiannon Ellis, and Elizabeth Loftus, "Make My Memory: How Advertising Can Change Our Memories of the Past," *Psychology and Marketing* 19 (2002): 1–23.

12. Barthes, *Camera Lucida*, 13.

13. Sontag, in *On Photography*, also touches upon this issue when she explains how retouching techniques were prominent in commercial portrait photography from the onset: "People want the idealized image: a photograph of themselves at their best. They feel rebuked when the camera does not return an image of themselves as more attractive than they really are" (85).

14. Jacques Aumont, in *The Image* (London: British Film Institute, 1997), provides a semiotic model that expands Barthes's theory of the photographic image to the contextual level, including the relationship between not only image and viewer but also image and its referent in terms of aesthetics and of representation. For a subtle critique of Barthes's paradoxes concerning his control and lack of control over the photographic image, see Ron Burnett, *Cultures of Vision: Images, Media and the Imaginary* (Bloomington: Indiana University Press, 1995), 32–71.

15. Strange, Gerrie, and Gary, "A Few Seemingly Harmless Routes to a False Memory," 237.

16. In recent years, there has been an explosion of theory on the semiotics and ontology of the digital image, but it is beyond the scope of this chapter to review the literature in this area. As a general introduction to the digitization of visual culture in general and photography in particular, consult *The Photographic Image in Digital Culture*, ed. Martin Lister (New York: Routledge, 1995) and Warwick Mules, "Lines, Dots and Pixels: The Making and Remaking of the Printed Image in Visual Culture," *Continuum, Journal of Media and Cultural Studies*, 14, no. 3 (2000): 303–16. A more philosophical introduction to ontology of the image can be found in D. N. Rodowick, *Reading the Figural, or, Philosophy after the New Media* (Durham, NC: Duke University Press, 2001). Lev Manovich, in *The Language of New Media* (Cambridge, MA: MIT Press, 2001), addresses digital photography as a technical-cultural construct. One notable exception to the negligence of cognitive perspectives by humanists is the work of Canadian art and design scholar Ron Burnett, who weaves cognitive perspectives into his semiotic approach to digital photography. In his recent book *How Images Think* (Cambridge, MA: MIT Press, 2004), Burnett scrutinizes mental-cognitive processes (with a specific emphasis on memory) in relation to digital technology.

17. William J. T. Mitchell, in *The Reconfigured Eye: Visual Truth in the Postphotographic Era* (Cambridge, MA: MIT Press, 1992), identifies the pictorial tradition of realism with the essence of photographic technology, and he identifies the tradition of montage and collage with the essence of digital imaging. Lev Manovich, in his essay "The Paradoxes of Digital Photography," published in Photography after Photography, Exhibition catalogue, Germany 1995 (available at http://www.manovich.net/TEXT/digital_photo.html, accessed December 26, 2006), refutes this claim, countering that both traditions existed before photography and that "normal" or "straight" photography never existed.

18. For a more technical and institutional exploration of photography's move toward digitization, see Fred Ritchin, *In Our Own Image: The Coming Revolution in Photography* (New York: Aperture, 1999).

19. For an interesting introduction to issues of image manipulation and stock photography, see, for instance, Paul Frosh, *The Image Factory: Consumer Culture, Photography, and the Visual Content Industry* (Oxford: Berg, 2003). In chapter 7, Frosh quotes directors from stock photography firms, who estimate that in the

late 1990s, 80 percent to 90 percent of all photographs they promoted had been digitally manipulated. As Frosh concludes: "Stock photographers and their clients in design forms and advertising agencies assume that consumers, long accustomed to the formal conventions and promotional goals of advertising images, do not expect such fidelity from their photographs" (175).

20. See, for instance, the software offered by VisionQuest Images, available at: http://www.visionquestimages.com/index.htm (accessed April 8, 2006).

21. The photoblog of Chris Line, available at: http://a.trendyname.org/archives/category/personal/ (accessed April 8, 2006).

22. For more details on the work of Nancy Burson, see http://www.nancyburson.com/human_fr.html (accessed April 8, 2006). Her Human Race Machine featured in many magazines and television programs in the spring of 2006, most notably on the *Oprah Winfrey Show.*

23. For instance, a package called Picture Yourself Graphics (available at: http://www.pygraphics.com) encourages playful collages and manipulation of pictures on wedding announcements, featuring the bride and groom as five-year-olds holding hands; other software design favors the use of personal pictures in combination with snapshots of famous tourist sites. For software that encourages the mix of personal photographs with general stock photography, promoting playful personalization of tourist snapshots, see http://www.fotosearch.com/photodisc/picture-yourself-here (accessed April 8, 2006).

24. Burnett, *How Images Think,* 28.

25. For a more elaborate argument on how ultrasound helps to turn the fetus into an object to be worked on, see José van Dijck, *Manufacturing Babies and Public Consent: Debating the New Reproductive Technologies* (New York: New York University Press, 1995), chapter 5.

26. Sontag, in *On Photography,* already called attention to photography's materialness by pointing at the logic of consumption underpinning the need to photograph things and people, converting sights into tangible, mobile objects: "To consume means to burn up—and therefore, the need to be replenished" (179).

27. Don Slater, "Domestic Photography and Digital Culture," in *The Photographic Image in Digital Culture,* ed. Martin Lister (New York: Routledge: 1995), 129–46, 130.

28. Barthes, in his *Camera Lucida,* already noticed how we tend to look through the laminated object because only the referent adheres, but he stresses how it is chemistry and light that turn a picture into a fetish-object to be looked at and to hold on to (p. 80).

29. For an interesting anthropological perspective on the materiality and rituality of photographs, see Elizabeth Edwards, "Photographs as Objects of Memory," in *Material Memories,* ed. Markus Kwint, Christopher Breward, and Jeremy Aymsley (Oxford: Berg, 1999), 221–48; see also Chris Wright, "Material and Memory. Photography in the Western Solomon Islands," *Journal of Material Culture* 9, no. 1 (2004): 73–85.

30. For an interesting take on the material objects accompanying virtual desktops, see Anna McCarthy, "Cyberculture or material culture?" *Etnofoor* 15, no. 1 (2002): 47–63.

31. There are a number of articles, most written by computer engineers, that grapple with the (new) performative meanings of digital photo management. See for instance, Kerry Rodden and Kenneth Wood, "How Do People Manage Their Digital Photographs?" *Computer Human Interaction* 5, no. 1 (2003): 409–16. See also Nancy Van House, Marc Davis, and Yuri Takhteyev, "From 'What' to 'Why': The Social Uses of Personal Photos" (paper presented at the CSCW Conference in Chicago, November 6–10, 2004).

32. In reaction to Sontag's pejorative interpretation of the touristic photographic experience, Steve Garlick argues in "Revealing the Unseen: Tourism, Art, and Photography," *Cultural Studies* 16, no. 2 (2002): 289–305, that tourist photography is "a mechanism that leads the subject to engage with the world in a creative act that opens and re-opens spaces, each time people take or review their pictures" (296).

33. Sontag, *On Photography*, 8.

34. Quite a few cultural theorists and anthropologists have taken up Sontag's insights to write about photography and family. See, for instance, Marianne Hirsch, *Family Frames: Photography, Narrative, and Postmemory* (Cambridge, MA: Harvard University Press, 1997); P. Holland, "Introduction: History, Memory, and the Family Album," in *Family Snaps: The Meaning of Domestic Photography*, ed. J. Spence and P. Holland (London: Virago, 1991): 1–14; and Deborah Chambers, *Representing the Family* (London: Sage, 2001).

35. Barbara Harrison, in "Photographic Visions and Narrative Inquiry," *Narrative Inquiry* 12, no.1 (2002): 87–111, notices a recent shift in photography's social use: "Images that have a place in everyday life have become less bound up with memory or commemoration, but with forms of practice that are happening now. . . . Self-presentation rather than self-representation is more important in identity formation" (107). Testifying to these trends is the pin board, where private pictures are often combined with public ones and personal pictures are put on display.

36. Barbara Harrison, "Photographic Visions," 107.

37. According to the *New York Times*, American sales of digital cameras surpassed the sales of analog film cameras for the first time in March 2003. See Katie Hafner, "Recording Another Day in America, Aided by Digital Cameras," *New York Times*, May 12, 2003 www.nytimes.com.

38. Diane Schiano, Coreena Chen, and Ellen Isaacs, "How Teens Take, View, Share, and Store Photos," in *Proceedings of the Conference on Computer-Supported Co-operative Work* (New York: ACM, 2002). Interestingly, the researchers conclude that teenagers are less inclined than adults to label pictures with captions, either because they don't think it is relevant or because they put great trust in their future memory capacity; they feel confident they will always be able to remember what is in the pictures.

39. The American study is corroborated by a Japanese report identifying similar patterns among young users of digital cameras; a new preference for photography as an interpersonal tool for communication whose main function is to exchange "affective awareness" prompts hardware and software developers to redirect their frameworks for design. See Oliver Liechti and Tadao Ichikawa, "A Digital Photography Framework Enabling Affective Awareness in Home Communication," *Personal and Ubiquitous Computing* 4, no. 1 (2000): 6–24.

40. For an incisive ethnographic study of photoblogs, see Kris Cohen, "What Does the Photoblog Want?" *Media, Culture and Society* 27, no. 6 (2005): 883–901.

41. See, for instance, Tim Kindberg, Mirjana Spasojevic, Rowanne Fleck, and Abigail Sellen, "I Saw This and Thought of You: Some Social Uses of Camera Phones," *Conference on Human Factors in Computing Systems* (New York: ACM, 2005), 1545–48.

42. A field study by a group of Finnish researchers yields this interesting comparison of the pictures sent by mobile phone to postcards. See Turo-Kimmo Lehtonen, Ilpo Koskinen, and Esko Kurvinen, "Mobile Digital Pictures—The Future of the Postcard? Findings from an Experimental Field Study," in: *Postcards and Cultural Rituals*, ed. V. Laakso and J-O Ostman (Korttien Talo: Haemeenlinna, 2002), 69–96.

43. For ethnographic research on the use of camera phone pictures, see Nancy Van House, Marc Davis, and Morgan Ames, "The Uses of Personal Networked Digital Imaging: An Empirical Study of Cameraphone Photos and Sharing," *Conference on Human Factors in Computing Systems* (New York: ACM, 2005), 1853–56.

44. Joseph Pine and James Gilmore, *The Experience Economy: Work Is a Theatre and Every Business a Stage* (Cambridge, MA: Harvard Business School, 1999).

45. Ron Burnett, in *How Images Think*, uses the term "microcultures" to describe places "where people take control of the means of creation and production in order to make sense of their social and cultural experiences" (62). Like the invention and distribution of the Xerox machine gave rise to new methods of information dissemination, the digital camera allows new communicative and formative uses of photography.

46. The pictures were first made public in the press by journalist Seymour Hersh who wrote the article "Torture at Abu Ghraib: American Soldiers Brutalized Iraqis. How Far Up Does the Responsibility Go?" *New Yorker*, May 4, 2004, http://www.newyorker.com/.

47. Susan Sontag, "Regarding the Torture of Others," *New York Times Magazine*, May 23, 2004, 25–29.

48. As Marita Sturken has shown in "The Image as Memorial: Personal Photographs in Cultural Memory," in *The Familial Gaze*, ed. Marianne Hirsch (Hanover, NH: University Press of New England, 1999), 178–95, digital enhancement is part of the message's rhetoric; personal pictures are malleable and open to absorb an infinite number of public meanings, depending on the context in which

they appear: as victims of AIDS, casualties of war, or as missing children on milk cartons.

49. Of course, the loss of power over one's public image has long been a matter of debate in political circles; politicians' personal pictures have always been used and abused in election campaigns and political image making. See, for instance, Kiku Adatto, *Picture Perfect: The Art and Artifice of Public Image Making* (New York: HarperCollins, 1993).

CHAPTER 6

1. For an interesting exploration of the discussions concerning the construction of family in film and television, see *Shooting the Family: Cultural Values and Transnational Media,* ed. Patricia Pisters and Wim Staat (Amsterdam: Amsterdam University Press, 2005), particularly the introduction.

2. Antonio Damasio, *The Feeling of What Happens: Body and Emotion in the Making of Consciousness* (Orlando, FL: Harcourt, 1999), 11.

3. Gilles Deleuze, *Cinema 2: The Time Image* (Minneapolis: University of Minnesota Press, 2003). Originally published as *Cinema 2: L'image-temps* (Paris: Athlone Press, 1989).

4. There are a large number of books explaining the importance and meaning of Deleuze's philosophical concepts for film theory. See, for instance, Gregory Flaxman, ed., *The Brain is the Screen: Deleuze and the Philosophy of Cinema* (Minneapolis: University of Minnesota Press, 2000); Barbara M. Kennedy, *Deleuze and Cinema: The Aesthetics of Sensation* (Edinburgh: Edinburgh University Press, 2000); and Ronald Bogue, *Deleuze on Cinema* (New York: Routledge, 2003).

5. Patricia Pisters, in *The Matrix of Visual Culture: Working with Deleuze in Film Theory* (Stanford, CA: Stanford University Press, 2003), provides an insightful analysis of Deleuze's theory and film, particularly his conjecture that the apparatus of cinema and the dynamics of brain activity and neural patterns are fundamentally interlaced.

6. Deleuze, *Cinema 2: The Time Image,* 52.

7. For a detailed critique of Deleuze's work, see Mark B. Hansen, *New Philosophy for New Media* (Cambridge, MA: MIT Press, 2004). A more elaborate theory of "virtual embodiment" particularly in relation to virtual environments can be found in Mark B. Hansen, *Bodies in Code: Interfaces with Digital Media* (New York: Routledge, 2006).

8. Hansen, *New Philosophy for New Media,* 194.

9. One of my criticisms of Deleuze's theory concerns his disregard of movies as cultural forms and watching movies as a sociocultural practice. I am not saying, though, that Deleuze completely ignores culture and politics in his writings; the micro-politics of culture are discussed more generally in his works *A Thousand Plateaus: Capitalism and Schizophrenia* (London: Continuum, 1988) and, with Felix

Guattari, *Anti-Oedipus: Capitalism and Schizophrenia* (Minneapolis: University of Minnesota Press, 2003).

10. *The Matrix*, written and directed by Andy Wachowski and Larry Wachowski (Los Angeles: Warner Bros, 1999); *Strange Days* directed by Kathryn Bigelow, written by James Cameron (Los Angeles: Twentieth Century Fox, 1995).

11. Deleuze, *Cinema 2: The Time Image*, 262.

12. A second storyline in the movie narrates the protests of a violent anti-implant group who opposes any form of biotechnological recording; two activists chase Hakman in order to obtain the Bannister files, which would provide a damaging blow to the implant industry.

13. James M. Moran, *There's No Place Like Home Video* (Minneapolis: University of Minnesota Press, 2002).

14. Ibid., 59.

15. Ibid., 103.

16. Patricia Zimmerman, states in *Reel Families: A Social History of Amateur Film* (Bloomington: Indiana University Press, 1995): "Home movie-making, then, synchronized with the elevation of the nuclear family as the ideological center of all meaningful activity in the fifties" (134).

17. Besides Zimmerman, Richard Chalfen, in *Snapshots Versions of Life* (Bowling Green, OH: Bowling Green State University Press, 1987), also ignores the historically changing connection between technological substrates and sociocultural concepts of family.

18. Ozzie and Harriet, for instance, also formed a couple in real life. On the role of families on American television from the 1950s onward, see for instance Lynn Spigel, *Make Room for TV: Television and the Family Ideal in Postwar America* (Chicago: University of Chicago Press, 1992).

19. Contrary to popular belief, American networks have always produced portraits of dysfunctional families as a counterpoint to idealized family series. A number of series that featured family lives represented the struggles and conflicts inherent to the postwar generation raising families in middle class, suburban America. As George Lipsitz argues in *Time Passages: Collective Memory and American Popular Culture* (Minneapolis: University of Minnesota Press, 1990) in his chapter on family in early network television: "One might expect commercial television programs to ignore the problems of the nuclear family, to present an idyllic view of the commodity-centered life. But the industry's imperial ambition—the desire to have household watching at all times—encouraged exploitation of real fears and problems confronting viewers" (56). For a detailed history of nuclear and alternative American families on television, see also Ella Taylor, *Prime-Time Families: Television Cultures in Postwar America* (Berkeley: University of California Press, 1989).

20. Sean Cubitt, *Timeshift: On Video Culture* (London: Routledge, 1991), 37.

21. On the change of the nuclear family in the 1960s, see for instance Arlene Skolnick, *Embattled Paradise: The American Family in an Age of Uncertainty* (New York: Basic Books, 1991).

22. As Moran aptly observes in *There's No Place Like Home Video*, "Each medium attempts to provide a home audience's hankering for audiovisual images of themselves, borrowing from each other over time, thus inventing and reinventing each other's conventions of representation and patterns of interpersonal communication" (106).

23. On the new genre of family portrait and documentary techniques, see Jim Lane, *The Autobiographical Documentary in America* (Madison: University of Wisconsin Press, 2002), 94–95.

24. Jeffrey Ruoff, *An American Family: A Televised Life* (Minneapolis: University of Minnesota Press, 2002), xvii.

25. The series was exhaustively reviewed in the press when it aired in 1973 and afterward; there have also been numerous academic and scholarly articles written that deal with *An American Family* and its documentary mode. For an overview, I refer to the extensive bibliography included in Ruoff's *An American Family*.

26. Ruoff, *An American Family*, 29.

27. Ibid., 88.

28. All family members, but most notably Pat Loud and her oldest son Lance, conceded in hindsight that the presence of a film crew in their house forever changed family life, even decades after the series was aired. Pat Loud wrote a book on her experiences and frequently appeared on television, including when the family was revisited by camera crews ten, fifteen, and twenty years after the actual shooting. Lance Loud, who became a filmmaker himself, even furbished the last episode in the series: WNET/PBS aired a production of his struggle with, and eventual succumbing to, HIV/AIDS in 2001.

29. On the use of webcams in the private sphere, see Sheila Murphy, "Lurking and Looking: Webcams and the Construction of Cybervisuality," in *Moving Images: From Edison to Webcam*, ed. John Fullerton and Astrid Söderbergh Widding (London: John Libbey, 2000), 173–80. See also Michele White, "Too Close to See: Men, Women, and Webcams," *New Media and Society* 5, no. 1 (2003): 7–28.

30. According to Australian film theorist Keith Beattie, in *Documentary Screens: Nonfiction Film and Television* (Hampshire: Palgrave Macmillan, 2004), the digital camcorder is "creating new visual styles that situate the viewer in an intimate relationship with the subject of autobiography" (105).

31. Moran, *There's No Place Like Home Video*, 47.

32. See the Jacobs Family Website, available at: http://jacobsusa.com/main/ (accessed April 13, 2006).

33. This peculiar manifestation of what Arild Fetveit, in "Reality TV in the Digital Era: A Paradox in Visual Culture?" in *Reality Squared: Televisual Discourse on the Real*, ed. James Friedman (New Brunswick, NJ: Rutgers University Press,

2002), 119–37, has called the "ambiguous coexistence of digital manipulation and 'reality footage' can be explained by a desire to reclaim a sense of reality that is virtually absent in any static collage of digital photos and texts" (130).

34. Jacobs Family Website, http://jacobsusa.com/main/.

35. *The Osbournes* was produced and broadcast by MTV. Subsequent DVD versions of the various seasons' series were distributed by Miramax Home Entertainment. Not the entire Osbourne family participates in the series: one of their three children refused to appear on television. The first season premiered in 2003.

36. Although real life usually does not present itself thematically, each twenty-minute episode of *The Osbournes* is organized as a fast-paced edited sequence capitalizing on a single theme, such as "like father, like daughter" or "won't you be my neighbor."

37. For an early discussion on the construction of reality in documentary, see Bill Nichols, *Representing Reality: Issues and Concepts in Documentary* (Bloomington: Indiana University Press, 1991). In postmodern television culture, the conventions of reality are increasingly informed by standards of so-called reality TV, where a contrived family—whether ten people living in a Big Brother house or a group convening on a deserted island—is captured on a reality show by ubiquitous cameras. The legacy of home movie and video is still at work in documentary, just as the legacy of documentary is still at work in reality TV. In coming to terms with these new forms of reality on the screen, media theorist John Corner has coined the term "postdocumentary" to describe new forms of tele-factuality. See John Corner, "Performing the Real: Documentary Diversions," *Television and New Media* 3 (2002): 255–69.

38. Moran, in *There's No Place Like Home Video*, appropriately describes this difference: "Digital icons undermine the authority of the video image and distance the artist from the actual process of image creation: whereas analogue video's aesthetic has been valued as immediate, literal, and naturalistic, digitized video is more often construed as contrived, synthetic, and analytic" (13).

39. The documentary *Capturing the Friedmans*, directed by Andrew Jarecki (New York: Magnolia Pictures) came out in 2003. In that same year, the DVD containing two discs with the documentary and lots of extra materials was distributed by HBO Video.

40. While David had bought and frequently used the video camera, his brother Jesse had a habit of audio-taping the family's rows at the dining table. Some of these audio-taped fights can be heard in the documentary.

41. In the interview with Andrew Jarecki featured on the DVD, the director relates how in the middle of shooting the film, David Friedman came up with twenty-five hours of taped home video and consented to its being used for the documentary.

42. Andrew Jarecki, in an interview also included on the DVD, states that by making this film, he took an explicit stance against currently reigning notions of reality TV: "We are so tuned now to reality on television, but you watch the

Friedmans for a minute and you see the incredible power of 'real' reality as opposed to that reality on CBS-TV where you see people starving on an island, but you know all the while they are surrounded by cameras."

43. In this quote, Deleuze echoes Antoni Artaud's beliefs in cinema and the world. See Deleuze, *Cinema 2: The Time-Image*, 167. For an extensive explanation and commentary on this observation, see Patricia Pisters, *From Eye to Brain. Gilles Deleuze: Refiguring the Subject in Film Theory* (Ph-Diss, University of Amsterdam, 1999), 76–85.

44. Hansen, *New Philosophy for New Media*, 270.

45. Kyle Veale has meticulously described this phenomenon in "Online Memorialisation: The Web as a Collective Memorial Landscape for Remembering the Dead" *Fibreculture* 3. Online journal available at: http://journal.fibreculture.org/issue3/issue3_veale.html (accessed December 28, 2006).

46. Life on Tape, a Dutch producer, specializes in memorial videos, available at: http://www.lifeontape.nl (accessed December 28, 2006). Precious Memories and More, an American company, offers memorial DVDs to "help survivors memorialize their loved ones." Available at: http://www.preciousmemoriesandmore.com (accessed December 28, 2006).

CHAPTER 7

1. See Scott McQuire, *Visions of Modernity: Representations, Memory, Time and Space in the Age of the Camera* (London: Sage, 1998), 127–38.

2. Gottfried Wilhelm Leibnitz (1646–1716), a German philosopher and mathematician, was the inventor of differential calculus; he wanted to design a universal language, which would facilitate communication via a network of universities. Leibnitz's cylindrical computer, though never built, signified an important step forward from dead mechanical calculations to a flexible "ars combinatoria," which would differentiate between the feeding in of data and the calculation itself. Leibniz also philosophized about a computer based on a binary numerical system. Charles Babbage (1791–1871), a British engineer, is known as the "Father of Computing" for his contributions to the basic design of the computer through his "analytical engine." His previous "difference engine" was a special-purpose device intended for the production of tables. He never turned his prototypes into working devices.

3. Vannevar Bush (1890–1974) was president of the Carnegie Institute in Washington, DC (1939) and chair of National Advisory Committee for Aeronautics (1939) before he became the director of the Office of Scientific Research and Development. This last role was a presidential appointment and made him responsible for the six thousand scientists involved in the WWII effort.

4. Vannevar Bush's famous, canonized article "As We May Think" first appeared in *Atlantic Monthly*, July 1945, 101–8. Available at: http://www.theatlantic.com/doc/194507/bush, accessed December 30, 2006.

5. Bush, *As We May Think*, section 3 (no page numbers in electronic version).

6. Ibid., section 6.

7. For an inventory of critical assessments of Bush's fantasies, see Andreas Kitzmann, "Pioneer Spirits and the Lure of Technology: Vannevar Bush's Desk, Theodor Nelson's World," *Configurations*, 9 (2001): 441–59.

8. See Charlie Gere, "Brains-in-Vats, Giant Brains and World brains: The Brain as Metaphor in Digital Culture," *Studies in the History and Philosophy of Biological and Biomedical Sciences* 35 (2004): 351–66.

9. Hartmut Winkler, "Discourses, Schemata, Technology, Monuments: Outline for a Theory of Cultural Continuity," *Configurations* 10 (2002): 91–109, 103.

10. Canadian media theorist Michelle Kendrick, in "Interactive Technology and the Remediation of the Subject of Writing," *Configurations* 9 (2001): 231–51, has critically analyzed how notions of hypertext are structured analogously to the mind, promoting the "connection between a technology of links and nodes and the presumed associative ability of the mind" (231).

11. Timothy Mills, David Pye, David Sinclair, and Kenneth R. Wood, "Shoebox: A Digital Photo Management System," http://www.uk.research.att.com/dart/shoebox/(accessed September 2004). (The Shoebox website is no longer available.) Shoebox is part of a larger project, DART (Digital Asset Retrieval Technology) concerned with the management of digital media such as text and hypertext documents, images, audio and video recordings.

12. Image-based indexing is a technique to mechanically recognize images based on rough drawings by the user. For instance, if the user draws a girl in a dress and the software will search for that particular figure.

13. Ernst van Alphen, in "Symptoms of Discursivity: Experience, Memory, and Trauma," in *Acts of Memory: Cultural Recall in the Present,* ed. Mieke Bal, Jonathan Crew, and Leo Spitzer (Hanover, NH: University Press of New England, 1999), 24–40, distinguishes narrative memory to underline the discursive basis of experiences; human memory is inherently mediated and discursive, and it is exactly this quality that helps people make sense out of experience.

14. Mills, Pye, Sinclair, and Wood, "Shoebox." Although the Shoebox website is no longer available, the conclusion is shared by follow-up research initiated by Microsoft engineers in cooperation with the Computer Science department of the University of Cambridge. See Kerry Rodden and Kenneth Wood, "How Do People Manage Their Digital Photographs?" *Computer Human Interaction* 5, no. 1 (2003): 409–16.

15. Molly Stevens, Gregory Abowd, Khai Truong, and Florian Vollmer, "Getting into the Living Memory Box: Family Archives and Holistic Design," *Personal Ubiquitous Computing* 7 (2003): 210–16, 212.

16. Ibid., 213.

17. Peculiarly, designers of the Living Memory Box acknowledge the cyclic function of memory objects in a family's lifetime, but the various stages involved

in this cycle of memories—collect, relate, create, donate—somehow remain alien to the box's design.

18. See Nicholas Carriero, Scott Fertig, Eric Freeman, and David Gelernter, "The Lifestreams Approach to Reorganizing the Information World," http://www.cs.yale.edu/homes/freeman/lifestreams.html (accessed April 14, 2006).

19. Ibid.

20. This observation is shared by Chris Locke in "Digital Memory and the Problem of Forgetting," in *Memory and Methodology*, ed. Susannah Radstone (Oxford: Berg, 2000), 25–36, who explains: "Vannevar Bush's Memex has ceased to be a technological system that supplements human memory. Instead it has given birth to a system which is coming to characterize contemporary human memory. Ironically, a system that was intended to aid the fallible, forgetful human memory has instead become a metaphor for postmodern definitions of human memory itself" (35).

21. The scientific premises of the MyLifeBits project can be found in Jim Gemmell, Gordon Bell, Roger Lueder, Steven Drucker, and Curtis Wong, "MyLifeBits: Fulfilling the Memex Vision," *ACM Multimedia*, December 2002, http://research.microsoft.com/barc/MediaPresence/MyLifeBits.aspx (accessed April 14, 2006). Various other articles on the MyLifeBits project are also available at this website.

22. See, for instance, Julia Scheeres, "Saving Your Bits for Posterity," *Wired News*, December 6, 2002, htttp://www.wired.com/ (accessed April 20, 2006); Arthur Hissey, "Your Life—On the Web," Computer Research and Technology, company website available at: http://www.crt.net.au/etopics/mylifebits.htm (accessed April 20, 2006).

23. I took this quote from the news article "Software Aims to Put Your Life on a Disk, *New Scientist*, November 20, 2002, http://www.newscientist.com (accessed December 27, 2006).

24. Scheeres, "Saving Your Bits for Posterity."

25. N. Katherine Hayles, *Writing Machines* (Cambridge, MA: MIT Press, 2002).

26. Google's algorithms will of course never be inculcated in human memory, but the fact that Google recently acquired Blogger.com (one of the first start-ups to design weblog software) indicates the company has a sharp eye for the individual's potential to navigate the baffling complexity of public and private crossroads paving the digital world.

27. There is already an interesting body of research on the cultural meaning of material types of collecting and storing them for later remembrance. On the functions and practices of collecting, see for instance Brenda Danet, "Books, Letters, Documents: The Changing Aesthetics of Texts in Late Print Culture," *Journal of Material Culture* 2 (1997): 5–38; Werner Muensterberger, *Collecting: An Unruly Passion* (Princeton, NJ: Princeton University Press, 1994); and Susan

Pearce, *On Collecting: An Investigation into Collecting in the European Tradition* (New York and London: Routledge, 1999).

CHAPTER 8

1. See, for instance, some marketers of preformatted scrapbooks (digital and paper) such as Creative Memories (available at: http://www.creativememories.com, accessed December 30, 2006) whose mission statement says: "Creative Memories believes in and teaches the importance of preserving the past, enriching the present, and inspiring hope for the future. We promote the tradition of historian/storyteller and the importance of memory preservation and journaling for future generations. We offer quality, photo-safe scrapbook and album-making products and information."

2. Elisabeth Loftus, "Make-Believe Memories," *American Psychologist* (2003): 867–73, 872.

3. Cockpits of Starfighter jets or surgical units help students train their skills in handling harrowing situations that involve all possible senses: sight, touch, smell, sound.

4. See, for instance, Francesco Vincelli, "From Imagination to Virtual Reality: The Future of Clinical Psychology," *CyberPsychology and Behavior* 2, no. 3 (1999): 241–48. See also Barbara Rotbaum and Larry Hodges, "A Controlled Study of Virtual Reality Exposure Therapy for the Fear of Flying," *Journal of Consulting and Clinical Psychology* 68, no. 6 (2000): 1020–26.

Bibliography

Adatto, Kiku. *Picture Perfect: The Art and Artifice of Public Image Making.* New York: HarperCollins, 1993.

Alphen, Ernst van. "Symptoms of Discursivity: Experience, Memory, and Trauma" In *Acts of Memory: Cultural Recall in the Present*, edited by Mieke Bal, Jonathan Crew, and Leo Spitzer, 24–38. Hanover, NH: University Press of New England, 1999.

Assmann, Jan, and John Czaplicka. "Collective Memory and Cultural Identity." *New German Critique* 65 (1995): 125–33.

Aumont, Jacques. *The Image.* London: British Film Institute, 1997.

Auner, Joseph. "Making Old Machines Speak: Images of Technology in Recent Music," *ECHO: A Music-Centered Journal* 2, no. 2 (2000). http://www.humnet.ucla.edu/echo, accessed December 30, 2006.

Bal, Mieke, Jonathan Crew, and Leo Spitzer, eds. *Acts of Memory: Cultural Recall in the Present.* Hanover, NH: University Press of New England, 1999.

Balch, W. R. and B. S. Lewis. "Threads of Music in the Tapestry of Memory." *Memory and Cognition* 21 (1996): 21–28.

Barnet, Belinda. "Pack-Rat or Amnesiac? Memory, the Archive and the Birth of the Internet." *Continuum: Journal of Media and Cultural Studies* 15, no. 2 (2001): 217–31.

———. "The Erasure of Technology in Cultural Critique." *Fibreculture* 1 (2003). http://journal.fibreculture.org/, accessed December 30, 2006.

Barthes, Roland. *Camera Lucida: Reflections on Photography.* New York: Hill and Wang, 1981.

Baumgartner, H. "Remembrances of Things Past: Music, Autobiographical Memory, and Emotion." *Advances in Consumer Research* 19 (1992): 613–20.

Beattie, Keith. *Documentary Screens: Nonfiction Film and Television.* Hampshire: Palgrave Macmillan, 2004.

Benjamin, Walter. *One-Way Street and Other Writings.* London: Verso, 1979.

Bergson, Henri. *Matter and Memory.* London: George Allen and Unwin, 1911.

Bjarkman, Kim. "To Have and to Hold: The Video Collector's Relationship with an Eternal medium." *Television and New Media* 5, no. 3 (2004): 217–46.

Blanchard, Anita. "Blogs as Virtual Communities: Identifying a Sense of Community in the Julie/Julia Project." In *Into the Blogoshphere: Rhetoric, Community and Culture of Weblogs,* edited by Laura Gurak, Smiljana Antonijevio, Laurie Johnson, Clanoy Ratliff, and Jessica Reyman. Minneapolis: University of Minnesota, 2004. http://blog.lib.umn.edu/blogosphere/, accessed December 30, 2006.

Bluck, Susan. "Autobiographical Memory: Exploring Its Functions in Everyday Life." *Memory* 11, no. 2 (2003): 113–23.

Bluck, Susan, and Linda Levine. "Reminiscence as Autobiographical Memory: A Catalyst for Reminiscence Theory Development." *Ageing and Society* 18 (1998): 185–208.

Bogue, Ronald. *Deleuze on Cinema.* New York: Routledge, 2003.

Bolter, Jay D., and Richard Grusin. *Remediation: Understanding New Media.* Cambridge, MA: MIT Press, 1999.

Bourdieu, Pierre. *Outline of a Theory of Practice.* Cambridge: Cambridge University Press, 1977.

Bourtchouladze, Rusiko. *Memories Are Made of This: How Memory Works in Humans and Animals.* New York: Columbia University Press, 2002.

Bower, Gordon, and Joseph Forgas. "Affect, Memory, and Social Cognition." In *Cognition and Emotion,* edited by Joseph Forgas, Eric Eich, and Gordon Bower, 87–168. Oxford: Oxford University Press, 2000.

Braun, Kathryn, Rhiannon Ellis, and Elizabeth Loftus. "Make My Memory: How Advertising Can Change Our Memories of the Past." *Psychology and Marketing* 19 (2002): 1–23.

Bull, Michael. *Sounding Out the City: Personal Stereos and the Management of Everyday Life.* Oxford: Berg, 2000.

Burnett, Ron. *Cultures of Vision: Images, Media and the Imaginary.* Bloomington: Indiana University Press, 1995.

———. *How Images Think.* Cambridge, MA: MIT Press, 2004.

Bush, Vannevar. "As We May Think." *Atlantic Monthly* (July 1945): 101–8. http://www.theatlantic.com/doc/194507/bush, accessed December 30, 2006.

Caldwell, John. "Prime-Time Fiction Theorizes the Docu-real." In *Reality Squared: Televisual Discourse on the Real,* edited by James Friedman, 259–91. New Brunswick, NJ: Rutgers University Press, 2002.

Certeau, Michel de. *The Practice of Everyday Life.* Berkeley: University of California Press, 1984.

Chalfen, Richard. *Snapshots Versions of Life.* Bowling Green, OH: Bowling Green State University Press, 1987.

———. "Snapshots 'R' Us: The Evidentiary Problematic of Home Media." *Visual Studies* 17, no. 2 (2002): 141–49.

Chalfen, Richard, and Mai Murui. "Print Club Photography in Japan: Framing Social Relationships." *Visual Sociology* 16, no. 1 (2001): 55–73.

Chambers, Deborah. *Representing the Family.* London: Sage, 2001.

Chaney, David. *Cultural Change and Everyday Life.* Houndmills, UK: Palgrave, 2002.

Cheung, Charles. "A Home on the Web: Presentations of Self on Personal Home-pages." In *Web.Studies: Rewiring Media Studies for the Digital Age,* edited by David Gauntlett, 43–51. London: Arnold, 2000.

Clark, Andy. *Mindware: An Introduction to the Philosophy of Cognitive Science.* Oxford: Oxford University Press, 2001.

Cohen, Kris. "What Does the Photoblog Want?" *Media, Culture and Society* 27, no. 6 (2005): 883–901.

Connell, John, and Chris Gibson. *Sound Tracks: Popular Music, Identity and Place.* London: Routledge, 2003.

Connerton, Paul. *How Societies Remember.* Cambridge: Cambridge University Press, 1989.

Corner, John. "Performing the Real: Documentary Diversions." *Television and New Media* 3 (2002): 255–69.

Cubitt, Sean. *Timeshift: On Video Culture.* London: Routledge, 1991.

Damasio, Antonio. *The Feeling of What Happens: Body and Emotion in the Making of Consciousness.* Orlando, FL: Harcourt, 1999.

———. *Looking for Spinoza: Joy, Sorrow, and the Feeling Brain.* Orlando, FL: Harcourt, 2003.

Danet, Brenda. "Books, Letters, Documents: The Changing Aesthetics of Texts in Late Print Culture." *Journal of Material Culture* 2 (1997): 5–38.

Decker, William M. *Epistolary Practices: Letter Writing in America before Telecommunications.* Raleigh: University of North Carolina Press, 1998.

Deleuze, Gilles. *A Thousand Plateaus: Capitalism and Schizophrenia.* London: Continuum, 1988.

———. *Cinema 2: The Time Image.* Minneapolis: University of Minnesota Press, 2003.

Deleuze, Gilles, and Felix Guattari. *Anti-Oedipus: Capitalism and Schizophrenia.* Minneapolis: University of Minnesota Press, 2003.

Derrida, Jacques. *Archive Fever: A Freudian Impression.* Chicago: University of Chicago Press, 1995.

Didier, Beatrice. *Le journal intime.* Paris: Editions du Seuil, 1976.

Dijck, José van. *Manufacturing Babies and Public Consent: Debating the New Reproductive Technologies.* New York: New York University Press, 1995.

———. *ImagEnation: Popular Images of Genetics.* New York: New York University Press, 1998.

———. "No Images without Words." In *The Image Society: Essays on Visual Culture,* edited by Frits Gierstberg and Warna Oosterbaan, 36–47. Rotterdam: Nai Publishers, 2002.

———. *The Transparent Body: A Cultural Analysis of Medical Imaging.* Seattle: University of Washington Press, 2005.

Draaisma, Douwe. *Metaphors of Memory: A History of Ideas about the Mind.* Cambridge: Cambridge University Press, 2000.

——. *Waarom de tijd sneller gaat als je ouder wordt.* Groningen: Historische Uitgeverij, 2001.

Dumit, Joseph. "Objective Brains, Prejudicial Images." *Science in Context* 12, no. 1 (1999): 173–201.

Edwards, Elizabeth. "Photographs as Objects of Memory." In *Material Memories,* edited by Markus Kwint, Christopher Breward, and Jeremy Aymsley, 221–48. Oxford: Berg, 1999.

Elsaesser, Thomas. " 'Where were you when . . . ?'; or, 'I Phone, therefore I am.' " *PMLA* 118, no. 1 (2003): 120–22.

Fernback, Jan. "Legends on the Net: An Examination of Computer-Mediated Communication as a Locus of Oral Culture." *New Media and Society* 5, no. 1 (2003): 29–45.

Fetveit, Arild. "Reality TV in the Digital Era: A Paradox in Visual Culture?" In *Reality Squared: Televisual Discourse on the Real,* edited by James Friedman, 119–37. New Brunswick, NJ: Rutgers University Press, 2002.

Flaxman, Gregory, ed. *The Brain Is the Screen: Deleuze and the Philosophy of Cinema.* Minneapolis: University of Minnesota Press, 2000.

Foucault, Michel. *The Archaeology of Knowledge and the Discourse on Language.* New York: Pantheon, 1972.

Frith, Simon. *Performing Rites: On the Value of Popular Music.* Oxford: Oxford University Press, 1996.

Frosh, Paul. *The Image Factory: Consumer Culture, Photography, and the Visual Content Industry.* Oxford: Berg, 2003.

Garland, Brent, ed. *Neuroscience and the Law: Brain, Mind, and the Scales of Justice.* New York: Dana Press, 2004.

Garlick, Steve. "Revealing the Unseen: Tourism, Art, and Photography." *Cultural Studies* 16, no. 2 (2002): 289–305.

Garry, Maryanne, and Matthew Gerrie. "When Photographs Create False Memories." *Current Directions in Psychological Science* 14 (2005): 321.

Garry, Maryanne, and Kimberley Wade. "Actually, a Picture Is Worth Less Than 45 Words: Narratives Produce More False Memories Than Photographs Do." *Psychonomic Bulletin and Review* 12, no. 2 (2005): 359–66.

Gere, Charlie. "Brains-in-Vats, Giant Brains and World Brains: The Brain as Metaphor in Digital Culture." *Studies in the History and Philosophy of Biological and Biomedical Sciences* 35 (2004): 351–66.

Gibbs, Anna. "Contagious Feelings: Pauline Hanson and the Epidemiology of Affect." *Australian Humanities Review* 24 (2001). http://www.lib.latrobe.edu.au/AHR/archive/Issue-December-2001/gibbs.html, accessed December 30, 2006.

Gitelman, Lisa. "How Users Define New Media: A History of the Amusement Phonograph." In *Rethinking Media Change: The Aesthetics of Transition,* edited by David Thorburn and Henry Jenkins, 61–80. Cambridge, MA: MIT Press, 2003.

Goodwin, Alan. "Sample and Hold: Pop Music in the Digital Age of Reproduction." *Critical Quarterly* 30, no. 3 (1988): 34–49.

Grainge, Paul. "Nostalgia and Style in Retro America: Moods, Modes, and Media Recycling." *Journal of American and Comparative Cultures* 23, no. 1 (2000): 27–34.

Gross, David. *Lost Time: On Remembering and Forgetting in Late Modern Culture.* Amherst: University of Massachusetts Press, 2000.

Halbwachs, Maurice. *On Collective Memory.* Chicago: University of Chicago Press, 1992.

Hansen, Mark. *Embodying Technesis: Technology beyond Writing.* Ann Arbor: University of Michigan Press, 2000.

———. *New Philosophy for New Media.* Cambridge MA: MIT Press, 2004.

———. *Bodies in Code. Interfaces with Digital Media.* New York: Routledge, 2006.

Harrison, Barbara. "Photographic Visions and Narrative Inquiry." *Narrative Inquiry* 12, no. 1 (2002): 87–111.

Hayles, N. Katherine. *How We Became Posthuman: Virtual Bodies in Cybernetics, Literature and Informatics.* Chicago: University of Chicago Press, 1999.

———. *Writing Machines.* Cambridge, MA: MIT Press, 2002.

Hirsch, Marianne. *Family Frames: Photography, Narrative, and Postmemory.* Cambridge, MA: Harvard University Press, 1997.

Hogan, Patrick Colm. *Cognitive Science, Literature, and the Arts.* New York: Routledge, 2003.

Holland, P. "Introduction: History, Memory, and the Family Album." In *Family Snaps: The Meaning of Domestic Photography*, edited by J. Spence and P. Holland, 1–14. London: Virago, 1991.

Hoskins, Andrew. "New Memory: Mediating History." *Historical Journal of Film, Radio and Television* 21, no. 4 (2001): 333–46.

———. "Signs of the Holocaust: Exhibiting Memory in a Mediated Age." *Media, Culture and Society* 25, no. 1 (2003): 7–22.

Hutchins, Edwin. *Cognition in the Wild.* Cambridge, MA: MIT Press, 1996.

Huyssen, Andreas. *Twilight Memories: Marking Time in a Culture of Amnesia.* New York: Routledge, 1995.

Intraub, Helene, and James Hoffman. "Reading and Visual Memory: Remembering Scenes that Were Never Seen." *American Journal of Psychology* 105, no. 1 (1992): 101–14.

Johnson, Steven. *Mind Wide Open: Your Brain and the Neuroscience of Everyday Life.* New York: Scribner, 2004.

Jones, Steve. "Music that Moves: Popular Music, Distribution and Network Technologies." *Cultural Studies* 16, no. 2 (2002): 213–32.

Katz, Mark. *Capturing Sound: How Technology Has Changed Music.* Berkeley: University of California, 2004.

Kawaura, Yasuyuki, Yoshiro Kawakami, and Kiyomi Yamashita. "Keeping a Diary in Cyberspace." *Japanese Psychological Research* 40, no. 4 (1998): 234–45.

Keightley, Keir. "Long Play: Adult-Oriented Popular Music and the Temporal Logics of the Postwar Sound Recording Industry in the USA." *Media, Culture and Society* 26 (2004): 375–91.

Keller, Ianus, Pieter Jan Stappers, and Sander VroegindeWeij. "Supporting Informal Collections of Digital Images: Organizing, Browsing and Sharing." *Proceedings on Dutch Directions in Human-Computer Interfaces* (2004): 1–14.

Kendrick, Michelle. "Interactive Technology and the Remediation of the Subject of Writing." *Configurations* 9 (2001): 231–51.

Kennedy, Barbara M. *Deleuze and Cinema: The Aesthetics of Sensation.* Edinburgh: Edinburgh University Press, 2000.

Kennedy, Quinn, Mara Mather, and Laura Carstensen. "The Role of Motivations in the Age-Related Positivity Effect in Autobiographical Memory." *Psychological Science* 15, no. 3 (1994): 208–14.

Kenney, William. *Recorded Music in American Life: The Phonograph and Popular Memory, 1890–1945.* Oxford: Oxford University Press, 1999.

Ketelaar, Eric. "Sharing: Collected Memories in Communities of Records." *Archives and Manuscripts* 23 (2005): 44–61.

Kindberg, Tim, Mirjana Spasojevic, Rowanne Fleck, and Abigail Sellen. "I Saw This and Thought of You: Some Social Uses of Camera Phones." In *Conference on Human Factors in Computing Systems,* 1545–48. New York: ACM, 2005.

Kittler, Friedrich. *Film, Gramophone, Typewriter.* Stanford, CA: Stanford University Press, 1999.

Kitzmann, Andreas. "Pioneer Spirits and the Lure of Technology: Vannevar Bush's Desk, Theodor Nelson's World." *Configurations* 9 (2001): 441–59.

———. "That Different Place: Documenting the Self within Online Environments." *Biography* 26, no. 1 (2003): 48–65.

Kotarba, Joseph. "Rock 'n' Roll Music as a Timepiece." *Symbolic Interaction* 25, no. 3 (2002): 397–404.

Kuhn, Annette. *Family Secrets: Acts of Memory and Imagination.* London: Verso, 1995.

———. "A Journey through Memory." In *Memory and Methodology,* edited by Susannah Radstone, 183–96. Oxford: Berg, 2000.

Lakoff, George, and Mark Johnson. *Metaphors We Live By.* Chicago: Chicago University Press, 1980.

Landsberg, Alison. *Prosthetic Memory: The Transformation of American Remembrance in the Age of Mass Culture.* New York: Columbia University Press, 2004.

Lane, Jim. *The Autobiographical Documentary in America.* Madison: University of Wisconsin Press, 2002.

Latour, Bruno. *We Have Never Been Modern.* Cambridge, MA: Harvard University Press, 1993.

————. *Aramis, or the Love of Technology.* Cambridge, MA: Harvard University Press, 1996.

Le Goff, Jacques. *History and Memory.* New York: Columbia University Press, 1992.

Lehtonen, Turo-Kimmo, Ilpo Koskinen, and Esko Kurvinen. "Mobile Digital Pictures—The Future of the Postcard? Findings from an Experimental Field Study." In *Postcards and Cultural Rituals,* edited by V. Laakso and J. O. Ostman, 69–96. Korttien Talo: Haemeenlinna, 2002.

Lejeune, Philip. *Le pacte autobiographique.* Paris: Editions du Seuil, 1993.

Leslie, Esther. "Souvenirs and Forgetting: Walter Benjamin's Memory-Work." In *Material Memories,* edited by Marius Kwint, Christopher Breward, and Jeremy Aynsley, 107–23. Oxford: Berg, 1999.

Lessig, Lawrence. *The Future of Ideas: The Fate of the Commons in a Connected World.* New York: Vintage, 2002.

Levine, Linda J. "Reconstructing Memories for Emotions." *Journal of Experimental Psychology* 126 (1997): 176–67.

Liechti, Olivier, and Tadao Ichikawa. "A Digital Photography Framework Enabling Affective Awareness in Home Communication." *Personal and Ubiquitous Computing* 4, no. 1 (2000): 6–24.

Lindsay, Stephen, Lisa Hagen, Don Read, Kimberley Wade, and Maryanne Garry. "True Photographs and False Memories." *Psychological Science* 15, no. 3 (2004): 149.

Lipsitz, George. *Time Passages: Collective Memory and American Popular Culture.* Minneapolis: University of Minnesota Press, 1990.

Lister, Martin, ed. *The Photographic Image in Digital Culture.* New York: Routledge, 1995.

Locke, Chris. "Digital Memory and the Problem of Forgetting." In *Memory and Methodology,* edited by Susannah Radstone, 25–36. Oxford: Berg, 2000.

Loftus, Elisabeth. "The Reality of Repressed Memories." *American Psychologist* 48 (1993): 518–37.

————. "Make-Believe Memories." *American Psychologist* (2003): 867–73.

Loftus, Elisabeth, and J. Pickrell. "The Formation of False Memories." *Psychiatric Annals* 25 (1995): 720–25.

Mallon, Thomas. *A Book of One's Own: People and Their Diaries.* New York: Ticknor and Fields, 1984.

Manovich, Lev. *The Language of New Media.* Cambridge, MA: MIT Press, 2001.

Martin, L. H. *Technologies of the Self: A Seminar with Michel Foucault.* London: Tavistock, 1988.

Marty, Eric. *L'ecriture du jour: Le journal d'André Gide.* Paris: Editions du Seuil, 1985.

Mather, Mara. "Aging and Emotional Memory." In *Memory and Emotion,* edited by Daniel Reisberg and Paula Hertel, 272–307. Oxford: Oxford University Press, 2004.

McCarthy, Anna. "Cyberculture or Material Culture?" *Etnofoor* 15, no. 1 (2002): 47–63.

McLuhan, Marshall. *Understanding Media: The Extensions of Man.* New York: McGraw Hill, 1964.

McQuire, Scott. *Visions of Modernity: Representations, Memory, Time and Space in the Age of the Camera.* London: Sage, 1998.

Meyer, Leonard B. *Emotion and Meaning in Music.* Chicago: Chicago University Press, 1961.

Miller, Carolyn R., and Dawn Shepherd. "Blogging as Social Action: A Genre Analysis of the Weblog." In *Into the Blogoshphere: Rhetoric, Community and Culture of Weblogs,* edited by Laura Gurak, Smiljana Antonijevio, Laurie Johnson, Clanoy Ratliff, and Jessica Reyman. Minneapolis: University of Minnesota, 2004. http://blog.lib.umn.edu/blogosphere/, accessed December 30, 2006.

Milne, Esther. "Email and Epistolary Technologies: Presence, Intimacy, Disembodiment." *Fibreculture* 2 (2004). http://journal.fibreculture.org/issue2/issue2 _milne.html, accessed December 30, 2006.

Misztal, Barbara. *Theories of Social Remembering.* Maidenhead, UK: Open University Press, 2003.

Mitchell, William J. T. *The Reconfigured Eye: Visual Truth in the Post-photographic Era.* Cambridge, MA: MIT Press, 1992.

Moran, James M. *There's No Place Like Home Video.* Minneapolis: University of Minnesota Press, 2002.

Morton, David. *Off the Record: The Technology and Culture of Sound Recording in America.* New Brunswick, NJ: Rutgers University Press, 2000.

Muensterberger, Werner. *Collecting: An Unruly Passion.* Princeton, NJ: Princeton University Press, 1994.

Mules, Warwick. "Lines, Dots and Pixels: The Making and Remaking of the Printed Image in Visual Culture." *Continuum: Journal of Media and Cultural Studies* 14, no. 3 (2000): 303–16.

Murphy, Sheila C. "Lurking and Looking: Webcams and the Construction of Cybervisuality." In *Moving Images: From Edison to Webcam,* edited by John Fullerton and Astrid Söderbergh Widding, 173–80. London: John Libbey, 2000.

Neef, Sonja. "Die (rechte) Schrift und die (linke) Hand." *Kodikas/Ars Semiotica* 25, no. 1 (2002): 159–76.

———. "Authentic Events. The Diaries of Anne Frank and the Alleged Diaries of Adolf Hitler." In *Sign Here! Handwriting in the Age of Technological Reproduction,* edited by Sonja Neef, José van Dijck, and Eric Ketelaar, 23–50. Amsterdam: Amsterdam University Press, 2006.

Neef, Sonja, José van Dijck, and Eric Ketelaar, eds. *Sign Here! Handwriting in the Age of Technological Reproduction.* Amsterdam: Amsterdam University Press, 2006.

Nelson, Katherine. "Narrative and Self, Myth and Memory: Emergence of the Cultural Self." In *Autobiographical Memory and the Construction of a Narrative Self: Developmental and Cultural Perspectives,* edited by Robyn Fivush and Catherine A. Haden, 3–25. Mahwah, NJ: Lawrence Erlbaum, 2003.

————. "Self and Social Functions: Individual Autobiographical Memory and Collective Narrative." *Memory* 11, no. 2 (2003): 125–36.

Nichols, Bill. *Representing Reality: Issues and Concepts in Documentary*. Bloomington: Indiana University Press, 1991.

Nora, Pierre. "Between Memory and History: Les lieux de memoire." *Representations* 26 (1989): 69–85.

Nora, Tia de. *Music in Everyday Life*. Cambridge: Cambridge University Press, 2000.

O'Brien, Geoffrey. *Sonata for Jukebox: Pop Music, Memory, and the Imagined Life*. New York: Counterpoint, 2004.

Olick, Jeffrey K. and Joyce Robbins. "Social Memory Studies: From Collective Memory to the Historical Sociology of Mnemonic Practices." *Annual Review of Sociology* 24 (1998): 105–40.

Ong, Walter J. *Orality and Literacy: The Technologizing of the Word*. London: Routledge, 1982.

Payne, Robert. "Digital Memories, Analogues of Affect." *Scan: Journal of Media Arts Culture* 2 (2004). http://scan.net.au/scan/journal/display.php?journal_id=42, accessed December 30, 2006.

Pearce, Susan. *On Collecting: An Investigation into Collecting in the European Tradition*. New York: Routledge, 1999.

Piggott, Michael. "Towards a History of the Australian Diary." In *Proceedings of I-Chora Conference*. International Conference on the History of Records and Archives, 68–75. Toronto: University of Toronto, 2003.

Pine, Joseph, and James Gilmore. *The Experience Economy: Work Is a Theatre and Every Business a Stage*. Cambridge, MA: Harvard Business School, 1999.

Pisters, Patricia. *The Matrix of Visual Culture: Working with Deleuze in Film Theory*. Stanford, CA: Stanford University Press, 2003.

Pisters, Patricia, and Wim Staat, eds. *Shooting the Family: Cultural Values and Transnational Media*. Amsterdam: Amsterdam University Press, 2005.

Prager, Jeffrey. *Presenting the Past: Psychoanalysis and the Sociology of Misremembering*. Cambridge, MA: Harvard University Press, 1998.

Radstone, Susannah, ed. *Memory and Methodology*. Oxford: Berg, 2000.

Ritchin, Fred. *In Our Own Image: The Coming Revolution in Photography*. New York: Aperture, 1999.

Rodden, Kerry, and Kenneth Wood. "How Do People Manage Their Digital Photographs?" *Computer Human Interaction* 5, no. 1 (2003): 409–16.

Rodowick, D. N. *Reading the Figural, or, Philosophy after the New Media*. Durham, NC: Duke University Press, 2001.

Rodzvilla, John, ed. *We've Got Blog: How Weblogs Are Changing Our Culture*. Cambridge, MA: Perseus, 2002.

Rose, Steven. *The Making of Memory*. London: Bantam Press, 1992.

Rotbaum, Barbara, and Larry Hodges. "A Controlled Study of Virtual Reality Exposure Therapy for the Fear of Flying." *Journal of Consulting and Clinical Psychology* 68, no. 6 (2000): 1020–26.

Rothenbuhler, Eric W. "Commercial Radio as Communication." *Journal of Communication* 46 (1996): 125–43.

Rothenbuhler, Eric W. and John D. Peters, "Defining Phonography: An Experiment in Theory." *Musical Quarterly* 81, no. 2 (1997): 242–64.

Ruoff, Jeffrey. *An American Family: A Televised Life.* Minneapolis: University of Minnesota Press, 2002.

Samuel, Raphael. *Theatres of Memory.* Volume 1: *Past and Present in Contemporary Culture.* London: Verso, 1994.

Schaap, Frank. "Links, Lives, Logs: Presentation in the Dutch Blogosphere." In *Into the Blogoshphere: Rhetoric, Community and Culture of Weblogs,* edited by Laura Gurak, Smiljana Antonijevio, Laurie Johnson, Clanoy Ratliff, and Jessica Reyman. Minneapolis: University of Minnesota, 2004. http://blog.lib.umn.edu/blogosphere/, accessed December 30, 2006.

Schiano, Diane J., Coreena Chen, and Ellen Isaacs. "How Teens Take, View, Share, and Store Photos." In *Proceedings of the Conference on Computer-Supported Co-operative Work.* New York: ACM, 2002.

Schulkind, Matthew D., Laura K. Hennis, and David C. Rubin. "Music, Emotion and Autobiographical Memory: They're Playing Your Song." *Memory and Cognition* 27, no. 6 (1999): 948–55.

Seremetakis, C. Nadia, ed. *The Senses Still: Perception and Memory as Material Culture in Modernity.* Boulder, CO: Westview Press, 1994.

Silverstone, Roger, Eric Hirsch, and David Morley. "Information and Communication Technologies and the Moral Economy of the Household." In *Consuming Technologies: Media and Information in Domestic Spaces,* edited by Roger Silverstone and Eric Hirsch, 14–31. London: Routledge, 1992.

Skolnick, Arlene. *Embattled Paradise: The American Family in an Age of Uncertainty.* New York: Basic Books, 1991.

Slater, Don. "Domestic Photography and Digital Culture." In *The Photographic Image in Digital Culture,* edited by Martin Lister, 129–46. New York: Routledge, 1995.

Sontag, Susan. *On Photography.* New York: Delta, 1973.

Spigel, Lynn. *Make Room for TV: Television and the Family Ideal in Postwar America.* Chicago: University of Chicago Press, 1992.

Stephens, Mitchell. *The Rise of the Image, the Fall of the Word.* New York: Oxford University Press, 1998.

Sterne, Jonathan. *The Audible Past: Cultural Origins of Sound Reproduction.* Durham, NC: Duke University Press, 2003.

———. "MP3 as Cultural Artifact." *New Media and Society,* 8 no. 5 (2006): 825–42.

Stevens, Molly, Gregory Abowd, Khai Truong, and Florian Vollmer, "Getting into the Living Memory Box: Family Archives and Holistic Design." *Personal Ubiquitous Computing* 7 (2003): 210–16.

Strange, Deryn, Matthew Gerrie, and Maryanne Garry. "A Few Seemingly Harmless Routes to a False Memory." *Cognitive Process* 6 (2005): 237–42.

Stuhlmiller, C. M. "Narrative Picturing: Ushering Experiential Recall." *Nursing Inquiry* 3: (1996): 183–84.

Sturken, Marita. *Tangled Memories: The Vietnam War, The AIDS Epidemic, and the Politics of Remembering.* Berkeley: University of California Press, 1997.

———. "The Image as Memorial: Personal Photographs in Cultural Memory." In *The Familial Gaze*, edited by Marianne Hirsch, 178–95. Hanover, NH: University Press of New England, 1999.

Sutton, John. *Philosophy and Memory Traces: Descartes to Connectionism.* Cambridge: Cambridge University Press, 1998.

———. "Porous Memory and the Cognitive Life of Things." In *Prefiguring Cyberculture: An Intellectual History*, edited by Darren Tofts, Annemarie Johnson, and Alessio Cavallara, 130–41. Cambridge, MA: MIT Press, 2002.

Tacchi, Joe. "Radio Texture: Between Self and Others." In *Material Cultures: Why Some Things Matter*, edited by Daniel Miller, 25–45. London: UCL Press, 1998.

Taylor, Alex S., and Richard Harper. "The Gift of the Gab? A Design-Oriented Sociology of Young People's Use of Mobiles." *Journal of Computer Supported Cooperative Work* 12, no. 3 (2003): 267–96.

Taylor, Ella. *Prime-Time Families: Television Cultures in Postwar America.* Berkeley: University of California Press, 1989.

Taylor, Timothy D. *Sounds: Music, Technologies, and Culture.* New York: Routledge, 2001.

Thacker, Eugene. "What is Biomedia?" *Configurations* 11 (2003): 47–79.

Thompson, John B. *The Media and Modernity: A Social Theory of the Media.* Cambridge UK: Polity Press, 1995.

Thorburn, David, and Henry Jenkins, eds. *Rethinking Media Change: The Aesthetics of Transition.* Cambridge, MA: MIT Press, 2003.

Tofts, Darren, Annemarie Johnson, and Alessio Cavallara, eds. *Prefiguring Cyberculture: An Intellectual History.* Cambridge, MA: MIT Press, 2002.

Tomkins, Silvan. *Affect, Imagery, Consciousness.* New York: Springer, 1962.

Turino, Thomas. "Signs of Imagination, Identity, and Experience: A Peircian Semiotic Theory for Music." *Ethnomusicology* 43, no. 2 (1999): 221–55.

Ulmer, Gregory. *Heuretics: The Logic of Invention.* Baltimore, MD: Johns Hopkins University Press, 1994.

Urry, John. "How Societies Remember the Past." In *Theorizing Museums: Representing Identity and Diversity in a Changing World*, edited by Sharon Macdonald and Gordon Fyfe, 45–68. Oxford: Blackwell, 1996.

Uttal, William R. *The New Phrenology.* Cambridge, MA: MIT Press, 2003.

Van House, Nancy, Marc Davis, and Morgan Ames. "The Uses of Personal Networked Digital Imaging: An Empirical Study of Cameraphone Photos and Sharing." In *Conference on Human Factors in Computing Systems*, 1853–36. New York: ACM, 2005.

Veale, Kylie. "Online Memorialisation: The Web as a Collective Memorial Landscape for Remembering the Dead." *Fibreculture* 3 (2004). http://journal.fibreculture.org/issue3/issue3_veale.html, accessed December 30, 2006.

Vincelli, Francesco. "From Imagination to Virtual Reality: The Future of Clinical Psychology." *CyberPsychology and Behavior* 2, no. 3 (1999): 241–48.

Wade, Kimberley, Maryanne Garry, Don Read, and Stephen Lindsay. "A Picture Is Worth a Thousand Lies: Using False Photographs to Create False Childhood Memories." *Psychonomic Bulletin and Review* 9, no. 3 (2002): 597–603.

Wang, Qi, and Jens Brockmeier. "Autobiographical Remembering as Cultural Practice: Understanding the Interplay between Memory, Self and Culture." *Culture and Psychology* 8, no. 1 (2002): 45–64.

White, Michele. "Too Close to See: Men, Women, and Webcams." *New Media and Society* 5, no. 1 (2003): 7–28.

Williams, Raymond. *Television: Technology and Cultural Form.* Hanover, NH: University of New England Press, 1974.

Willis, Paul. *Common Culture: Symbolic Work at Play in the Everyday Cultures of Young.* London: Open University Press, 1990.

Winkler, Hartmut. "Discourses, Schemata, Technology, Monuments: Outline for a Theory of Cultural Continuity." *Configurations* 10 (2002): 91–109.

Wright, Chris. "Material and Memory. Photography in the Western Solomon Islands." *Journal of Material Culture* 9, no. 1 (2004): 73–85.

Yakel, Elizabeth. "Reading, Reporting, and Remembering: A Case Study of the Maryknoll Sisters Diaries." In *Proceedings of I-Chora Conference*, 142–50. Toronto: University of Toronto, 2003.

Zak, Albin J. *The Poetics of Rock: Cutting Tracks, Making Records.* Berkeley: University of California Press, 2001.

Zaltman, Gerald. *How Customers Think: Essential Insights into the Mind of the Market.* Boston: Harvard Business School Press, 2003.

Zhang, Jane. "The Lingering of Handwritten Records." In *Proceedings of I-Chora Conference*, 38–45. Toronto: University of Toronto, 2003.

Zimmerman, Patricia R. *Reel Families: A Social History of Amateur Film.* Bloomington: Indiana University Press, 1995.

Index

Abu Ghraib, 116–117, 120
Acrobat, The, 170
Adventures of Ozzie and Harriet, The, 133
AIDS Memorial Quilt, 22
All in the Family, 134, 136
Alphen, Ernst van, 214n13
Alzheimer's disease (AD), 55, 58–60, 74,
   161; patients of, 59–62, 68, 72, 75, 168,
   171
American Family, An, 134–136, 211n28
American Office of Scientific Research
   and Development, 150
America's Funniest Home Videos (AFHV),
   19, 184n11
Anne Frank Foundation, 63
Artaud, Antoni, 213n43
"As We May Think," 150
Assmann, Aleida, 12, 14, 22, 185n36
Assmann, Jan, 12
AT&T Labs, 154
Augustine, 29
Aumont, Jacques, 205n14
Auner, Joseph, 89

Babbage, Charles, 150–151, 213n2
Barnet, Belinda, 26
Barthes, Roland, 100–103, 106, 206n28
Beattie, Keith, 211n30
Bell, Gordon, 159, 165–166
Benjamin, Walter, 36
Bergson, Henri, 9, 29–30, 125–126
biomedia, 45
Blogger, 65

Bluck, Susan, 3, 4
Bourdieu, Pierre, 190n24
Bourtchouladze, Rusiko, 200n12
Brady Bunch, The, 134
Braun, Katryn, 204n11
Brockmeier, Jens, 4
Bull, Michael, 92
Burnett, Ron, 106, 205n16, 208n45
Burson, Nancy, 106, 180, 206n22
Bush, Vannevar, 149–154, 158–159, 213n3

camera: Easy Share, 111; Instamatic, 111;
   Kodak's Brownie, 111; Polaroid, 111
camera phone, 110, 114–115, 176
Capturing the Friedmans, 140–147
Carstensen, Laura, 200n19
Certeau, Michel de, 190n24
Chalfen, Richard, 17
Chaney, David, 197n35
cinema verité, 135
cinema-of-the-brain, 127
cinematic hindsight, 128, 131, 144, 147
Clark, Andy, 38
commons, creative, 97
Connell, John, 84
Corner, John, 212n37
Creative Memories, 216n1
Cubitt, Sean, 133
Cyclops Camera, 151

Damasio, Antonio, 34, 35, 81–82, 86,
   125–126, 189n10
DearDiary, 67

Decker, William, 196n31
Deleuze, Gilles, 123, 125–128, 131, 139, 145, 147, 209n9
Dementia and Alzheimer's Support Network International (DASNI), 58, 194n11
Derrida, Jacques, 64
Descartes, René, 29
Diary of Anne Frank, The, 58
DiaryLand, 65
Didier, Beatrice, 194n8
digitization, 42–45, 49, 108, 162–163
direct cinema. *See* cinema verité
Draaisma, Douwe, 17, 187n42
Durkheim, Emile, 9

Elsaesser, Thomas, 21
England, Lynndie, 116
Eternal Sunshine of the Spotless Mind, 27–34, 37, 40, 44–46, 51, 128

family film. *See* home movie
family movie. *See* home movie
Fernback, Jan, 197n36
Fetveit, Arild, 211n33
Final Cut, The, 128–130, 147
fly-on-the-wall cinema. *See* cinema verité
Foucault, Michel, 39
Frank, Anne, 13, 57, 62–63, 68–69
Frank, Otto, 13
Freud, Sigmund, 64, 158
Friedell, Morris, 58, 60–61, 194n11
Frith, Simon, 200n17
Frosh, Paul, 205n19

Garlick, Steve, 207n32
Garry, Maryanne, 204n9
genomics, 192n36
Georgia Institute of Technology, 156
Gerber, Chip, 59, 61
Gerrie, Matthew, 204n9
Getty Foundation, 105
Gibbs, Anna, 56, 59
Gibson, Chris, 84
Gide, André, 68

Gilbert, Craig, 135
Gilmore, James, 115
Gitelman, Lisa, 80
Gondry, Michel, 28, 44
Goodwin, Alan, 202n31
Google, 166, 215n26
googlization, 150, 162, 167, 177
Grainge, Paul, 96, 203n42
Graner, Charles, 116
Gross, David, 10, 12, 184n17

habitus, 190n24
Halbwachs, Maurice, 9–12
Hansen, Mark B., 123, 127–128, 130–131, 139, 146–147, 209n7
Harper, Richard, 197n41, 198n43
Harrison, Barbara, 113, 207n35
Hayles, N. Katherine, 161, 191n33
Hennis, Laura, 200n21
Herring, Susan, 193n1
Hirsch, Eric, 41
Hogan, Patrick Colm, 80
Holocaust, 11, 13
home mode, 131–132, 136–137, 140–141, 146
home movie, 123, 128–130, 133, 135, 141, 144–146
home video, 131, 133, 135; digital, 139–141, 144–146, 175
Hoskins, Andrew, 11, 12, 186n28
Human Race Machine, 106, 206n22
Hume, David, 29
Hutchins, Edwin, 191n26
Huyssen, Andreas, 11, 185n22

image: action, 126; affection, 126; movement, 126; perception, 126
Imperial War Museum, 11
Instamatic. *See* camera

Jarecki, Andrew, 141–142, 212n41
Johnson, Steven, 32

Katz, Mark, 90, 201n26
Kaufman, Charlie, 28
Kendrick, Michelle, 214n10

Kennedy, Quinn, 200n19
Kenney, William, 95
Kitzmann, Andreas, 193n4
Kotarba, Joseph, 201n23
Kuhn, Annette, 16, 184n9

Lacuna Inc., 27
Landsberg, Alison, 23
Latour, Bruno, 187n36
Le Goff, Jacques, 16
Leave It to Beaver, 133
Leibnitz, Gottfried, 150–151, 213n2
Lejeune, Philip, 194n8
Lessig, Lawrence, 97
Life on Tape, 147
lifelog, 53–54, 59–69, 73–75, 179
Lifestreams, 157–159, 161–162, 165, 167–68
Lipsitz, George, 17, 198n2, 210n19
LiveJournal, 66, 70, 76
Living Memory Box, The, 154–157,
    160–161, 165–168
Locke, Chris, 215n20
Lockhardt, Mary, 60

Magnetic Resonance Imaging (MRI),
    27; functional (f MRI), 27, 42, 45, 58,
    191n29
Mallon, Thomas, 67, 193n8
Manovich, Lev, 205n17
Marty, Eric, 194n8
Mather, Mara, 200n19
Matrix, The, 128
Matter and Memory, 30, 125
McLuhan, Marshall, 15
McQuire, Scott, 150
media: digital 26, 178–179; home 18, 19;
    mass 16–19; morphing, 175,
Memento, 32–33, 128
memex, 149, 151–153
memories: embedded, 28, 38, 79, 90; em-
    bodied, 28–29, 38, 79, 123; enabled,
    28, 35, 38–41; mediated, 2, 21–25, 28,
    38–39, 49–52, 55, 94, 126, 136, 145, 164,
    173–174; 180, 182
memory: acts of, 6, 7, 13, 47, 49, 127, 168,

171; autobiographical, 2–5, 9, 18, 31–32,
56, 77–78, 81, 89, 99–101, 105, 119, 124,
162, 169, 172, 179; collective cultural,
8–14, 16, 19, 22, 25, 77–78, 82, 91, 94,
97, 99–100, 117; institutions, 10; mediation of, 15–21; objects, 28, 35, 37, 41, 51,
112, 168, 171; personal cultural, 1, 6–11,
13, 17, 19, 22, 25, 46–47, 78, 82, 94,
99–100, 121, 124, 148, 168, 172, 178; social, 10, 185n26
microcultures, 208n45
Microsoft, 105, 166–167
Microsoft Media Presence Lab, 159
Miller, Carolyn, 197n42
Milne, Esther, 71
mindware, 34–35, 42, 45, 189n9
Misztal, Barbra, 201n25
MIT MediaLab, 71, 198n46
Mitchell, William J. T., 205n17
modifiability, condition of, 107
Moran, James, 123, 131–132, 136–139, 147,
    210n22, 212n38
Morley, David, 41
Morton, David, 199n5
movies-in-the-brain, 125–126
multimediatization, 150, 162, 175, 177–178
MyLifeBits, 154, 159–161, 163, 165–168

Naim, Omar, 128, 130–131
Neef, Sonja, 63
Nelson, Katherine, 3, 4, 183n4, 185n18
Nora, Pierre, 16
Nora, Tia de, 91
"Note Upon the 'Mystic Writing Pad,'
    A," 64,
Nussbaum, Emily, 66, 70, 198n45

O'Brien, Geoffrey, 84
Olick, Jeffrey K., 185n26
Ong, Walter, 15, 186n30
OpenDiary, 65–67
Osbournes, The, 138, 212n35

Partridge Family, The, 134
Payne, Robert, 60

Peirce, Charles, 82
photoblogs, 114, 179
photography, personal, 98, 101, 106–108,
    110, 112–114, 117–120, 203n1; digital
    104–105, 109–110, 112, 115, 118–120;
    phone, 114–115
Picabia, Francis, 170–173
Picture Yourself Graphics, 206n23
Piggott, Michael, 196n33
Pine, Joseph, 115
Pisters, Patricia, 209n5
plasticity, condition of, 107
Plato, 15
Polaroid. *See* camera
Positron Emission Tomography (PET),
    42–43, 58, 189n16, 191n29
Precious Memories, 147
privacy, 72–74, 76, 168
private mobilization, 88, 201n28
punctum, 103

Random Access Memory, 60
Raymond, Alan, 135
Raymond, Susan, 135
reconsolidation theory, 32–33, 47, 174
remediation, 47, 49
Robbins, Joyce, 185n26
Rose, Steven, 17, 203n6
Rothenbuhler, Eric, 201n26, 201n29
Rubin, David, 200n21
Ruoff, Jeffrey, 135

Samuel, Raphael, 16
Schiano, Diane, 207n38
Schulkind, Matthew, 200n21
Seventh Symphony (Gustave Mahler), 31
Shepherd, Dawn, 197n42
Shoah Visual History Foundation,
    185n20
Shoebox, 154–156, 160–161, 166–168
Silverstone, Roger, 41
Slater, Don, 108
Sontag, Susan, 112, 116, 204n13, 206n26
Spielberg, Steven, 185n20
Stephens, Mitchell, 186n31

Sterne, Jonathan, 93, 202n38
Strange Days, 128
Stuhlmiller, C. M., 203n2
Sturken, Marita, 22, 188n49, 208n48
subjectivity, affective, 55–56, 59–61
Sutton, John, 29, 38

Tacchi, Joe, 201n29
Taylor, Alex, 197n41, 198n43
Taylor, Timothy, 199n5
technologies: of affect, 46, 72; of mem-
    ory, 127, 31; of self (Foucault), 39, 48,
    51, 72, 162, 190n24; of truth (Fou-
    cault) 39, 162
technostalgia, 86–87
Thacker, Eugene, 45
Thompson, John, 18, 25, 187n45
Time-Image, The, 125
Tomkins, Silvan, 56
Top 2000, The Dutch National, 78,
    80–82, 84, 87–97, 199n6
trauma, 180–181
Turino, Tomas, 79, 82, 85, 88, 199n8

Ulmer, Gregory, 192n39
Urry, John, 17

Vietnam Memorial, 22
VisionQuest Images, 106
Vocoder, 151

Wade, Kimberley, 204n9, 204n10
Wang, Qi, 4
Wells, H. G., 44
Williams, Raymond, 88
Winkler, Hartmut, 152, 190n25

Xanga, 66, 70, 76

Yakel, Elizabeth, 196n32
Yates, Francis, 186n30

Zhang, Jane, 195n21
Zimmerman, Patricia, 132
Zoe Eye Tech Implant, 128–129, 151

Cultural Memory in the Present

Diana Sorensen, *A Turbulent Decade Remembered: Scenes From the Latin American Sixties*

Bella Brodzki, *"Can These Bones Live?": Translation, Survival, and Cultural Memory*

Asja Szafraniec, *Beckett, Derrida, and the Event of Literature*

Sara Guyer, *Romanticism after Auschwitz*

Gerard Richter, *Thought-Images: Frankfurt School Writers' Reflections from Damaged Life*

Alison Ross, *The Aesthetic Steering of Philosophy*

Rodolphe Gasché, *The Honor of Thinking: Critique, Theory, Philosophy*

Brigitte Peucker, *The Material Image: Art and the Real in Film*

Natalie Melas, *All the Difference in the World*

Jonathan Culler, *The Literary in Theory*

Michael G. Levine, *The Belated Witness*

Jennifer A. Jordan, *Structures of Memory*

Christoph Menke, *Reflections of Equality*

Marlène Zarader, *The Unthought Debt: Heidegger and the Hebraic Heritage*

Jan Assmann, *Religion and Cultural Memory: Ten Studies*

David Scott and Charles Hirschkind, *Powers of the Secular Modern: Talal Asad and his Interlocutors*

Gyanendra Pandey, *Routine Violence: Nations, Fragments, Histories*

James Siegel, *Naming the Witch*

J. M. Bernstein, *Against Voluptuous Bodies: Late Modernism and the Meaning of Painting*

Theodore W. Jennings, Jr., *Reading Derrida / Thinking Paul: On Justice*

Richard Rorty and Eduardo Mendieta, *Take Care of Freedom and Truth Will Take Care of Itself: Interviews with Richard Rorty*

Jacques Derrida, *Paper Machine*

Renaud Barbaras, *Desire and Distance: Introduction to a Phenomenology of Perception*

Jill Bennett, *Empathic Vision: Affect, Trauma, and Contemporary Art*

Ban Wang, *Illuminations from the Past: Trauma, Memory, and History in Modern China*

James Phillips, *Heidegger's* Volk: *Between National Socialism and Poetry*

Frank Ankersmit, *Sublime Historical Experience*

István Rév, *Retroactive Justice: Prehistory of Post-Communism*

Paola Marrati, *Genesis and Trace: Derrida Reading Husserl and Heidegger*

Krzysztof Ziarek, *The Force of Art*

Marie-José Mondzain, *Image, Icon, Economy: The Byzantine Origins of the Contemporary Imaginary*

Cecilia Sjöholm, *The Antigone Complex: Ethics and the Invention of Feminine Desire*

Jacques Derrida and Elisabeth Roudinesco, *For What Tomorrow . . . : A Dialogue*

Elisabeth Weber, *Questioning Judaism: Interviews by Elisabeth Weber*

Jacques Derrida and Catherine Malabou, *Counterpath: Traveling with Jacques Derrida*

Martin Seel, *Aesthetics of Appearing*

Nanette Salomon, *Shifting Priorities: Gender and Genre in Seventeenth-Century Dutch Painting*

Jacob Taubes, *The Political Theology of Paul*

Jean-Luc Marion, *The Crossing of the Visible*

Eric Michaud, *The Cult of Art in Nazi Germany*

Anne Freadman, *The Machinery of Talk: Charles Peirce and the Sign Hypothesis*

Stanley Cavell, *Emerson's Transcendental Etudes*

Stuart McLean, *The Event and its Terrors: Ireland, Famine, Modernity*

Beate Rössler, ed., *Privacies: Philosophical Evaluations*

Bernard Faure, *Double Exposure: Cutting Across Buddhist and Western Discourses*

Alessia Ricciardi, *The Ends Of Mourning: Psychoanalysis, Literature, Film*

Alain Badiou, *Saint Paul: The Foundation of Universalism*

Gil Anidjar, *The Jew, the Arab: A History of the Enemy*

Jonathan Culler and Kevin Lamb, eds., *Just Being Difficult? Academic Writing in the Public Arena*

Jean-Luc Nancy, *A Finite Thinking*, edited by Simon Sparks

Theodor W. Adorno, *Can One Live after Auschwitz? A Philosophical Reader*, edited by Rolf Tiedemann

Patricia Pisters, *The Matrix of Visual Culture: Working with Deleuze in Film Theory*

Andreas Huyssen, *Present Pasts: Urban Palimpsests and the Politics of Memory*

Talal Asad, *Formations of the Secular: Christianity, Islam, Modernity*

Dorothea von Mücke, *The Rise of the Fantastic Tale*

Marc Redfield, *The Politics of Aesthetics: Nationalism, Gender, Romanticism*

Emmanuel Levinas, *On Escape*

Dan Zahavi, *Husserl's Phenomenology*

Rodolphe Gasché, *The Idea of Form: Rethinking Kant's Aesthetics*

Michael Naas, *Taking on the Tradition: Jacques Derrida and the Legacies of Deconstruction*

Herlinde Pauer-Studer, ed., *Constructions of Practical Reason: Interviews on Moral and Political Philosophy*

Jean-Luc Marion, *Being Given That: Toward a Phenomenology of Givenness*

Theodor W. Adorno and Max Horkheimer, *Dialectic of Enlightenment*

Ian Balfour, *The Rhetoric of Romantic Prophecy*

Martin Stokhof, *World and Life as One: Ethics and Ontology in Wittgenstein's Early Thought*

Gianni Vattimo, *Nietzsche: An Introduction*

Jacques Derrida, *Negotiations: Interventions and Interviews, 1971-1998*, ed. Elizabeth Rottenberg

Brett Levinson, *The Ends of Literature: The Latin American 'Boom" in the Neoliberal Marketplace*

Timothy J. Reiss, *Against Autonomy: Cultural Instruments, Mutualities, and the Fictive Imagination*

Hent de Vries and Samuel Weber, eds., *Religion and Media*

Niklas Luhmann, *Theories of Distinction: Re-Describing the Descriptions of Modernity*, ed. and introd. William Rasch

Johannes Fabian, *Anthropology with an Attitude: Critical Essays*

Michel Henry, *I am the Truth: Toward a Philosophy of Christianity*

Gil Anidjar, *"Our Place in Al-Andalus": Kabbalah, Philosophy, Literature in Arab-Jewish Letters*

Hélène Cixous and Jacques Derrida, *Veils*

F. R. Ankersmit, *Historical Representation*

F. R. Ankersmit, *Political Representation*

Elissa Marder, *Dead Time: Temporal Disorders in the Wake of Modernity (Baudelaire and Flaubert)*

Reinhart Koselleck, *The Practice of Conceptual History: Timing History, Spacing Concepts*

Niklas Luhmann, *The Reality of the Mass Media*

Hubert Damisch, *A Childhood Memory by Piero della Francesca*

Hubert Damisch, *A Theory of /Cloud/: Toward a History of Painting*

Jean-Luc Nancy, *The Speculative Remark: (One of Hegel's bon mots)*

Jean-François Lyotard, *Soundproof Room: Malraux's Anti-Aesthetics*

Jan Patočka, *Plato and Europe*

Hubert Damisch, *Skyline: The Narcissistic City*

Isabel Hoving, *In Praise of New Travelers: Reading Caribbean Migrant Women Writers*

Richard Rand, ed., *Futures: Of Jacques Derrida*

William Rasch, *Niklas Luhmann's Modernity: The Paradoxes of Differentiation*

Jacques Derrida and Anne Dufourmantelle, *Of Hospitality*

Jean-François Lyotard, *The Confession of Augustine*

Kaja Silverman, *World Spectators*

Samuel Weber, *Institution and Interpretation: Expanded Edition*

Jeffrey S. Librett, *The Rhetoric of Cultural Dialogue: Jews and Germans in the Epoch of Emancipation*

Ulrich Baer, *Remnants of Song: Trauma and the Experience of Modernity in Charles Baudelaire and Paul Celan*

Samuel C. Wheeler III, *Deconstruction as Analytic Philosophy*

David S. Ferris, *Silent Urns: Romanticism, Hellenism, Modernity*

Rodolphe Gasché, *Of Minimal Things: Studies on the Notion of Relation*

Sarah Winter, *Freud and the Institution of Psychoanalytic Knowledge*

Samuel Weber, *The Legend of Freud: Expanded Edition*

Aris Fioretos, ed., *The Solid Letter: Readings of Friedrich Hölderlin*

J. Hillis Miller / Manuel Asensi, *Black Holes / J. Hillis Miller; or, Boustrophedonic Reading*

Miryam Sas, *Fault Lines: Cultural Memory and Japanese Surrealism*

Peter Schwenger, *Fantasm and Fiction: On Textual Envisioning*

Didier Maleuvre, *Museum Memories: History, Technology, Art*

Jacques Derrida, *Monolingualism of the Other; or, The Prosthesis of Origin*

Andrew Baruch Wachtel, *Making a Nation, Breaking a Nation: Literature and Cultural Politics in Yugoslavia*

Niklas Luhmann, *Love as Passion: The Codification of Intimacy*

Mieke Bal, ed., *The Practice of Cultural Analysis: Exposing Interdisciplinary Interpretation*

Jacques Derrida and Gianni Vattimo, eds., *Religion*